Contents

PAUL C. FIFE
University of Utah

Dynamics of Internal Layers and Diffusive Interfaces

SOCIETY FOR INDUSTRIAL AND APPLIED MATHEMATICS
PHILADELPHIA, PENNSYLVANIA 1988

Library of Congress Catalog Card Number 88-60958.

ISBN 0-89871-225-4.

Typeset for the Society for Industrial and Applied Mathematics by The Universities Press, Belfast, Ireland. Printed by Eastern Lithographing Corporation, Philadelphia, Pennsylvania.

Preface

Interfacial phenomena are commonplace in physics, chemistry, biology, and in various disciplines bridging these fields. They occur whenever a continuum is present that can exist in at least two different chemical or physical "states," and there is some mechanism that generates or enforces a spatial separation between these states. The separation boundary is then called an interface. In the examples studied here, the separation boundary and its internal structure result from the balance between two opposing tendencies: a diffusive effect that attempts to mix and smooth the properties of the material and a physical or chemical mechanism that works to drive it to one or the other pure state.

This latter is an "unmixing" tendency. In our cases, it is one of the following: (1) a chemical kinetic mechanism with two stable steady states or two attracting slow manifolds in concentration space; (2) a double-well potential that drives a substance into one of two possible phases, such as solid or liquid; (3) an imposed electric field that affects different kinds of ions in different ways; (4) a chemical reaction rate that is so sensitive to temperature that a temperature isocline can serve as an interface separating (a) the region in space where the reaction (and other reactions it triggers) has gone to completion, from (b) the other region, where the reaction is so slow due to the lower temperature that it can be neglected; or finally (5) a very complex biophysical process responsible for the triggering of physical change in biological tissue, followed by its recovery to its original state.

This fifth mechanism, together with the diffusive-type process alluded to above, is responsible for the propagation of signals along a nerve axon or cardiac tissue, and is quite commonly modeled by systems of the type studied in Chap. 4. The Belousov–Zhabotinskii and other excitable chemical reagents subject to mechanism (1) above are appropriately modeled the same way, and are also treated in Chap. 4.

The separation of ions due to the electric field mentioned in (3) is explained in Chap. 3. The thermal propagation of flames, caused by mechanism (4) plus the heat release of the reaction and the diffusion of heat and material species, is the subject of Chap. 2. A phase change model with far-reaching consequences is considered in Chap. 1, §3, and in Chap. 1, §2, we explain an equation that has been used to study waves in population genetics, physiology, and nonlinear transmission lines, and is an essential ingredient for the analysis in Chap. 4. Separations due to a diffusive shock layer are briefly covered in Chap. 1, §4.

In view of all of this, it must be emphasized that this work is primarily mathematical. One of the reasons, in fact, for its existence is that a certain body of techniques and concepts forms the basis of the mathematical study of most of these types of interfaces. This common ground consists of the asymptotic study of internal layers, and is introduced in Chap. 1, §1.

There are important differences, however, among the applications treated here. In fact, the bulk of the monograph by far is devoted to problems and phenomena that are particular to the various contexts.

In all cases, the most interesting phenomena have to do with the motion of the interfaces. In the case of flame interfaces (called layers), the tendency for irregular motion leads to well-known stability considerations, as well as to an excellent example of the power of multiple scaling techniques for determining their motion near the onset of instability. The latter topic is only briefly mentioned in Chap. 2, §2, but the difficult linear stability analysis is covered in great detail there.

In the case of the chemical and biophysical interfaces studied in Chap. 4, their motion leads to such fascinating spatiotemporal structures as rotating spirals and expanding rings. Scroll rings, the three-dimensional analogues of spirals, are not covered, but conceptually are subject to the same considerations.

The basis of this monograph was a set of notes prepared for three minicourses which I gave in 1987: a CBMS Conference on Nonlinear Waves in Little Cottonwood Canyon, Utah, in May; a series of lectures at Peking University and the Institute of Mathematics, Beijing, also in May; and a Rocky Mountain Mathematical Consortium summer course in Laramie, Wyoming, in July. I am very grateful to the organizers of these courses: Peter Bates at Brigham Young University, Ye Qi-xiao and Hsiao Ling in Beijing, and Duane Porter at the University of Wyoming. The original notes were written while I was a Visiting Professor at Brigham Young University. I am also grateful to the many people who read parts of the notes and made corrections. Finally, thanks go to the National Science Foundation, which makes CBMS conferences such as the one in Utah possible.

Most of this material exclusive of Chap. 1, §1, Chap. 2, §§1,2, and Chap. 4, §4C, represents research supported by National Science Foundation grants DMS-8202056 and DMS-8703247, and Air Force Office of Scientific Research grant F4962086C0130.

CHAPTER 1

Internal Layers

1. Dynamics of internal layers: Asymptotics and matching.

Since most of the nonlinear waves considered in these lectures are interfaces in the form of internal layers, it is appropriate that we begin by discussing the latter in general terms. Asymptotic methods somewhat related to the ones given here appear in a number of books, such as [Ec].

Imagine some smooth state variable u, a function of space and time, taking values in R^m, and evolving according to some system of differential equations. For simplicity we will assume space to be two-dimensional. This dynamical process also depends, we suppose, on a small positive parameter ϵ, so u does as well: $u = u(x, t; \epsilon)$, $x = (x_1, x_2)$.

We will study several situations in which the dynamical process generates and preserves a moving internal layer, which we will call an interface, of width $O(\epsilon)$. The generation process is interesting and in most cases has been studied little; we overlook that part here and focus on the dynamics of a fully developed layer. This layer is located on a curve $\Gamma(t; \epsilon)$ in the plane (its exact relation to Γ will be made precise in each case). It will be convenient to track the movement of Γ by considering the evolution of the function $r(x, t; \epsilon)$ defined to be $\pm$ the distance from x to Γ. Thus

$$\Gamma(t; \epsilon) = \{r(x, t; \epsilon) = 0\}. \tag{1}$$

We suppose that Γ divides the plane into two parts: $\mathscr{D}_+$, where r is taken to be positive, and $\mathscr{D}_-$, where r is negative. Further assumptions are as follows.

The state function may be approximated by truncations of a formal power series

$$u(x, t) \simeq \Sigma \epsilon^n u_n(x, t) \tag{2}$$

in regions away from the interface, and of another formal power series near Γ. The higher the truncation, of course, the better the approximation. To be more specific, let $\Gamma_\delta(t; \epsilon) = \{|r| \le \delta\}$ and $\mathscr{D}_\delta(t; \epsilon)$ be the complement of Γ_δ in the plane. Let $u^{(N)}$ denote the summation on the right of (2) up to terms of order ϵ^N. Then for some $\delta(N, \epsilon)$ to be specified below, with $\delta \to 0$ as $\epsilon \to 0$, we assume

$$|u - u^{(N)}| = O(\epsilon^{N+1}) \quad \text{as } \epsilon \to 0$$

uniformly for $(x, t) \in \mathcal{D}_{\delta(N,\epsilon)}$. This is to be true for all N up to some integer N_0 that depends on the context. The same approximability relation is to hold for the corresponding derivatives of u and $u^{(N)}$ up to some order, again depending on the context.

The "outer" functions u_n may be discontinuous or otherwise nonsmooth on Γ, but are smooth in $\mathcal{D}_+$ uniformly up to Γ, and are also smooth in $\mathcal{D}_-$ uniformly up to Γ.

The representation of u near Γ is as follows. We consider a local orthogonal coordinate system $(r(x, t; \epsilon), s(x, t; \epsilon))$ in a neighborhood of $\Gamma(t; \epsilon)$. Here s, for x on Γ, represents arclength along Γ. We then introduce a stretched variable

$$\rho(x, t; \epsilon) \equiv \frac{r(x, t; \epsilon)}{\epsilon}, \tag{3}$$

and think of the same function u now expressed in terms of the coordinates ρ and s: $u(x, t; \epsilon) = U(\rho, s, t; \epsilon)$ in that neighborhood. Again, U and its derivatives are supposed to be approximable by truncations of a formal series

$$U \simeq \Sigma \epsilon^n U_n(\rho, s, t), \tag{4}$$

i.e., by polynomials in ϵ with coefficients depending only on ρ, s, and t. These are the "inner" approximations. To be more precise again, we suppose there is a function $K(\epsilon, N)$ with a property to be given later such that $K \to \infty$ as $\epsilon \to 0$, so that the approximation of U and its derivatives by the N-truncation of (4) has an error $\leq O(\epsilon^{N+1})$ uniformly for $|\rho| < K(\epsilon, N)$.

Denote the normal velocity $-(\partial r/\partial t)(s, t; \epsilon) = -r_t$ of the interface at the point s in the direction of positive r by the function $v(s, t; \epsilon)$. Assume that Γ depends uniformly smoothly on t and ϵ; it then follows that v and r can be approximated by truncated power series, uniformly in (s, t) in the case of v and uniformly in a neighborhood of Γ in the case of r. Note that knowledge of the function r determines s, in view of the orthogonality of the coordinate system and the fact that s measures arclength. The function s also may be approximated by a polynomial in ϵ.

We often will denote partial derivatives by ∂_t, etc. The symbol $\partial_{\rho\rho}$ will denote the second partial with respect to ρ.

The following relations between the outer functions u_i and the inner functions U_i are to hold. Their justification on the basis of the above assumptions for a particular choice of δ and K will be given below. They are called "matching conditions." In the following the outer functions will be written as functions of r and s rather than x. As $\rho \to \pm\infty$,

(5a) $$U_0(\pm\infty, s, t) = u_0(0\pm, s, t),$$

(5b) $$U_1(\rho, s, t) = u_1(0\pm, s, t) + \rho\partial_r u_0(0\pm, s, t) + o(1),$$

(5c) $$U_2(\rho, s, t) = u_2 + \rho\partial_r u_1 + \tfrac{1}{2}\rho^2\partial_{rr}u_0 + o(1),$$

etc. The arguments of the functions on the right of (5c) are the same as in (5b). Here the notation $\partial_r u_0(0\pm, s, t)$, for example, denotes the limit, as Γ is approached from $\mathcal{D}_\pm$ at the point s, of the normal derivative of the function u_0 in the direction of increasing r. Sometimes the symbol $\partial_r u_0(\Gamma\pm, t)$ will be used instead.

In each case studied here, the various coefficients U_n and u_n will be subject to determination from differential equations under suitable boundary conditions. The matching relations (5) will be instrumental in defining the desired solutions of these problems.

The various problems for the U_n and u_n will constitute formal reductions to various orders of refinement of our original evolution model for u, and in this sense will be alternate models for the natural phenomenon being described by that original problem.

One may object justifiably at this point to the fact that the approximability assumptions we have made in the inner and outer regions are very restrictive. How do we know a priori that this approach can be followed in any particular circumstance? Of course we do not, but experience indicates that it will be successful in many cases, including those studied here. Formal verification of these assumptions would be accomplished by being able to construct, reasonably and systematically, the various inner and outer functions. Rigorous verification would consist of proving that the approximations so constructed are indeed close to an exact solution of the original problem. This latter step is only rarely done; early examples of where it was done for stationary interfaces can be found in [FGr], [Fi74], [Fi76a], and [MTH]. Important recent results for quite general systems with internal layers were obtained by Lin [Lin]. The usual practice in the applied literature (at least implicitly) is to accept and interpret the problems for the various approximations as being the alternate models mentioned above.

Certain properties of the local coordinate system (r, s) are appropriately stated here, since they will be useful later on several occasions.

On Γ, we have

(6) $$|\nabla r| \equiv 1 \quad \text{and} \quad \Delta r = \kappa,$$

where Δ and ∇ refer only to the spatial variable x, and κ is the curvature of Γ, counted as positive if Γ is concave as seen from $\mathcal{D}_-$. The first equation in (6), in fact, holds in the entire neighborhood where r is defined.

A standard calculation (making use of (6) and the orthogonality of the local system) shows that the change to local variables near Γ transforms the

Laplacian and time derivative as follows:

$$\Delta u = \partial_{rr} u + \Delta r \partial_r u + \partial_s u \Delta s + \partial_{ss} u\, |\nabla s|^2, \tag{7a}$$

$$\partial_t u \quad \text{becomes} \quad \partial_t u + r_t \partial_r u + s_t \partial_s u. \tag{7b}$$

We now return to the justification of the matching relations (5). Suppose t is fixed; hence Γ is also. Near Γ, we write u as a function of the local coordinates r and s: $u = u(r, s; \epsilon)$. We equate u and U in that region:

$$U(\rho, s, t; \epsilon) = u(\epsilon\rho, s, t; \epsilon). \tag{8}$$

Both the inner and outer representations (2) and (4) will be valid in the sense explained above in an intermediate region $\mathcal{I}$ in the (ϵ, ρ) plane defined by

$$\mathcal{I}: \quad \epsilon^{-1}\delta(\epsilon, N) \le \rho \le K(\epsilon, N), \tag{9}$$

provided this intermediate region is nonempty, which will be guaranteed in a moment by our construction of it. Note that for fixed x, r and ρ also depend on ϵ; however, we are not fixing x, and so that dependence need not be considered at this stage. Now (8) still holds except for an error term of order $O(\epsilon^{N+1})$, if we replace U on the left by the truncation $U^{(N)}$ and u on the right by $u^{(N)}$. A similar statement holds for the derivatives of (8) up to some order. We make this replacement. Suppressing dependence on s and t, we have

$$U_0(\rho) + \cdots + \epsilon^N U_N(\rho) = u_0(\epsilon\rho) + \cdots + \epsilon^N u_N(\epsilon\rho) + O(\epsilon^{N+1}) \quad \text{in } \mathcal{I}. \tag{10}$$

Now since the u_i were assumed to be smooth up to Γ, we can expand them in finite Taylor series in r around $r = 0$, and thereby get an expansion (for $\rho > 0$) such as

$$u_0(r, s, t) = u_0(\epsilon\rho, s, t) = u_0(0+, s, t) + \epsilon(\rho\partial_r u_0(0+, s, t) + u_1(0+, s, t)) + \cdots.$$

There results

$$\sum_{n=0}^{N} \epsilon^n U_n(\rho, s, t) = \sum_{n=0}^{N} \epsilon^n P_n(\rho, s, t) + \epsilon^{N+1} R_N + O(\epsilon^{N+1}), \tag{11}$$

where

$$P_n(\rho, s, t) \equiv \frac{1}{n!}\frac{\partial^n}{\partial \epsilon^n} u^{(N)}(\epsilon\rho, s, t; \epsilon)\,\big|_{\epsilon=0}, \tag{12}$$

and R_N = (similar expression, with $n = N + 1$ and evaluated at some $\hat{\epsilon}$ rather than $\epsilon = 0$). (For $\rho > 0$, the expression on the right means the limit, as $\epsilon \downarrow 0$, of the derivative indicated. This is also true for $\rho < 0$.)

It is clear from (12) that for $\rho > 0$, P_n is a polynomial in ρ of degree n with coefficients depending (linearly) on the $u_k(0+, s, t)$, $\partial_r u_k(0+, s, t)$, $\partial_{rr} u_k(0+, s, t)$, etc. A similar statement holds when $\rho < 0$. Therefore, the functions P_n in general will be discontinuous at $\rho = 0$, and be polynomials on either side. To make this clear, on occasion we may write them as $P_n^{\pm}(\rho, s, t)$.

Moreover, $|R_N| \leq O(\rho^{N+1})$, so that

$$\sum_{n=0}^{N} \epsilon^n U_n(\rho, s, t) = \sum_{n=0}^{N} \epsilon^n P_n(\rho, s, t) + O(\epsilon^{N+1}\rho^{N+1}). \tag{13}$$

At this point we specify δ and K to be of the form

$$\delta = \epsilon^{\alpha}, \quad K = \epsilon^{-\beta}, \quad \beta > 1 - \alpha, \tag{14}$$

α and β being positive constants depending on N to be given below.

We will give the rest of the argument for the specific case $N = 2$; the extension to larger values of N will be immediate. Let $\mathcal{J}_\gamma$ denote the line $\epsilon = \rho^{-\gamma}$ in the (ϵ, ρ) plane, where γ is chosen so that $\mathcal{J}_\gamma \in \mathcal{J}$, in fact, so that

$$1 - \alpha < \frac{1}{\gamma} < \beta. \tag{15}$$

Because of the strict inequalities, for $\gamma' - \gamma$ small enough, $\mathcal{J}_{\gamma'} \in \mathcal{J}$ as well.

On $\mathcal{J}_\gamma$, (13) for $N = 2$ takes the form

$$(U_0 - P_0) + \rho^{-\gamma}(U_1 - P_1) + \rho^{-2\gamma}(U_2 - P_2) = O(\epsilon^3\rho^3) \quad \text{on } \mathcal{J}_\gamma. \tag{16}$$

Let the three terms on the left be denoted by V_0, V_1, and V_2. We may then write (16) in the form

$$\frac{V_0 + V_1 + V_2}{\epsilon^3\rho^3} \quad \text{is bounded on } \mathcal{J}_\gamma. \tag{17a}$$

Also let $\gamma' = \gamma + \eta$ for $\eta > 0$ sufficiently small so $\mathcal{J}_{\gamma'} \in \mathcal{J}$. Then the analogous relation is

$$\frac{V_0 + \rho^{-\eta}V_1 + \rho^{-2\eta}V_2}{\epsilon^3\rho^3\rho^{-3\eta}} \quad \text{is bounded on } \mathcal{J}_\gamma. \tag{17b}$$

In (17b) we may omit the factor $\rho^{-3\eta}$ in the denominator without changing the limit relation. We do this and subtract (17b) from (17a) to obtain

$$\frac{(1 - \rho^{-\eta})V_1(\rho) + (1 - \rho^{-2\eta})V_2(\rho)}{\epsilon^3\rho^3} \quad \text{is bounded on } \mathcal{J}_\gamma.$$

As $\rho \to \infty$, the powers of ρ indicated on the left approach zero, so we obtain as a consequence

$$V_1 + V_2 = O(\epsilon^3\rho^3) \quad \text{on } \mathcal{J}_\gamma \quad \text{as } \rho \to \infty, \tag{18}$$

and from (17a),

$$V_0 = O(\epsilon^3\rho^3) \quad \text{on } \mathcal{J}_\gamma \quad \text{as } \rho \to \infty. \tag{19}$$

Since the left side of (19) in fact does not depend on ϵ, we obtain

$$U_0(\rho) \to P_0 \quad \text{as } \rho \to \infty. \tag{20}$$

Now multiply (18) by $\rho^\gamma = \epsilon^{-1}$ to obtain

$$(U_1 - P_1) + \rho^{-\gamma}(U_2 - P_2) = O(\epsilon^2\rho^3) \quad \text{on } \mathcal{J}_\gamma \quad \text{as } \rho \to \infty.$$

Repeating the above argument, we get

$$U_1 - P_1 = O(\epsilon^2\rho^3) \quad \text{on } \mathcal{J}_\gamma \quad \text{as } \rho \to \infty. \tag{21}$$

Proceeding one more step, we get

$$U_2 - P_2 = O(\epsilon\rho^3) \quad \text{on } \mathcal{J}_\gamma \quad \text{as } \rho \to \infty. \tag{22}$$

It follows that if $0 < \beta < \frac{1}{3}$, then from (14) and (15) we have $\epsilon\rho^3 \to 0$ on $\mathcal{J}_\gamma$, and from (20), (21), (22),

$$U_n(\rho, s, t) = P_n^\pm(\rho, s, t) + o(1) \quad (\rho \to \pm\infty), \tag{23}$$

for $n = 0$, 1, and 2. (The superscript $\pm$ is inserted because clearly the same argument holds for $\rho < 0$.)

The same result holds for any N if we merely choose

$$0 < \beta < \frac{1}{N+1}.$$

And of course α must satisfy (15); this completes the specification of δ and K, and completes the assumptions under which we are operating.

In particular, it follows that the large ρ behavior of U^n must be that of a polynomial in ρ of degree $n-1$. The relations (5) give the specific results on this large ρ behavior for $n = 0$, 1, and 2. They were obtained by calculating P_0, P_1, and P_2 explicitly.

2. Example: Bistable fronts in an inhomogeneous medium.

Consider the single equation

$$\epsilon\partial_t u = \epsilon^2 \Delta u + f(u, x), \tag{24}$$

where f is of "bistable type" in u for each x. This means that for each x, the equation $f(u, x) = 0$ can be solved for exactly three values of u as functions $h(x)$, and the condition $f_u(h(x), x) < 0$ is satisfied at the two extremal functions (see Fig. 1.1). The maximal and minimal such functions are denoted by $h_+(x)$

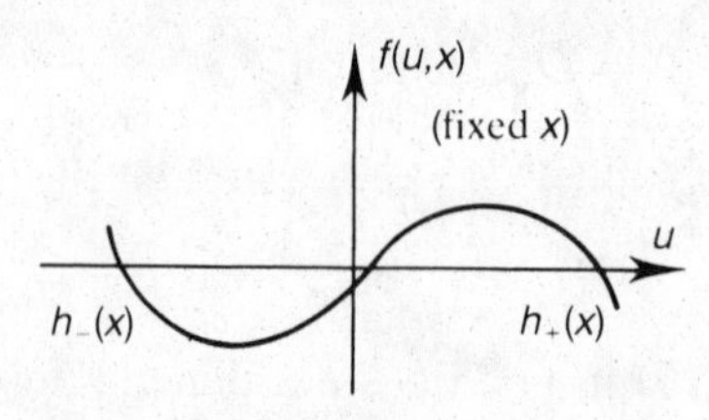

FIG. 1.1

and $h_-(x)$, respectively. For convenience, we suppose that

(25) $$h_-(x) < 0 < h_+(x).$$

Note that in (24) the presence of the coefficient ϵ of the time derivative is of no consequence. Time can be scaled so that any desired coefficient appears there; we choose ϵ for convenience. This is not true with the coefficient ϵ^2 of the Laplacian; in fact, rescaling x also changes the function f. That coefficient simply means that the ratio of the characteristic space scale of the solution in regions where diffusion is important to the characteristic scale of the function f's spatial variation is small, of order ϵ. In other words, relative to the spatial scale associated with diffusion, the function f varies slowly.

The theory of such bistable equations is very well developed in the case when f does not depend on x (then, of course, space, as well as time, may be rescaled to make all ϵ's disappear) and space is one-dimensional. We will make good use of those results. So let us fix $x = x_0$, and consider the corresponding equation

(26) $$\partial_t \psi = \partial_{xx} \psi + f(\psi, x_0).$$

It is known (see [Ka], [AW75], [AW78], [FM], [Fi79c], for example) that this equation has a globally stable traveling wave solution $\psi(x - \bar{v}t)$ satisfying $\psi(\pm\infty) = h_\pm(x_0)$ for exactly one value of $\bar{v}$. We will write it as $\psi(x - \bar{v}t, x_0)$ to indicate its dependence on x_0. Moreover, the solution ψ is unique, modulo shifts in the independent variable. Let us denote the velocity by the function

(27) $$\bar{v} = V(x_0).$$

This function is also known to be smooth if f is smooth [FH].

Now let us return to the original problem (24), with x in R^2. The outer functions u_n are obtained by substituting the formal series (2) into (24) and equating the coefficients of the various powers of ϵ. This ensures that the residual error in (24) produced by using the series truncated to order N is formally of order ϵ^{N+1}, which is the highest order possible.

The lowest order outer problem in this case is

(28) $$f(u_0, x) = 0.$$

As the solution of (28), we choose

(29) $$u_0(x) = h_-(x), \quad x \in \mathscr{D}_-; \quad u_0(x) = h_+(x), \quad x \in \mathscr{D}_+.$$

The solution of the inner problem will smooth out the discontinuity on Γ. Recall that in §1, the curve Γ was not defined precisely. To remedy this imprecision in the present case, we define it to be the location where the inner function U equals 0. Thus

(30) $$U(0, s, t) \equiv 0$$

to all orders.

For the inner problems we represent the time and space derivatives in (24) according to (7). Points on the moving interface $\Gamma(t)$ may be specified by coordinates (s, t); we denote them by the function $X(s, t)$. Points in a neighborhood of Γ, therefore, can be denoted by $X(s, t) + re_r$, where $e_r(s, t)$ denotes the unit vector normal to Γ at $X(s, t)$ in the direction of increasing r. We perform this transformation in (24) to obtain the following inner equation:

$$\text{(31)}\quad \begin{aligned} r_t\partial_\rho U + \epsilon\partial_t U + \epsilon s_t\partial_s U = \partial_{\rho\rho} U + f(U, X(s, t) + \epsilon\rho e_r) + \epsilon\Delta r\partial_\rho U \\ + \epsilon^2(\partial_{ss} U\,|\nabla s|^2 + \partial_s U\Delta s). \end{aligned}$$

Again, the inner functions U_n are obtained by substituting (4) into (31) and equating coefficients of the different powers of ϵ. To lowest order, we have (recalling $v_0 = -\partial_t r_0$)

$$\text{(32)}\quad \partial_{\rho\rho} U_0 + v_0\,\partial_\rho U_0 + f(U_0, X) = 0,$$

and the matching condition (5a) together with (29) requires

$$\text{(33)}\quad U_0(\pm\infty, s, t) = h_\pm(X(s, t)).$$

(This and (25) show that (30) may always be arranged.) Here and in (32), of course, $X = X(s, t)$ is any point on the interface. Now note that (32) is exactly the traveling wave equation for (26) with x_0 replaced by X, and in view of the uniqueness mentioned above, we must have

$$\text{(34)}\quad v_0 = V(X(s, t)),$$

$$\text{(35)}\quad U_0(\rho, s, t) = \psi(\rho, X(s, t)),$$

where ψ now denotes the solution of (26) with the independent variable shifted so that ψ vanishes at $\rho = 0$.

These facts suffice to obtain, to lowest order, the motion of Γ, because at each instant of time (34) gives the normal velocity of Γ at any point X on Γ. We therefore have the function $r_0(x, t)$ and hence $s_0(x, t)$. Along with that, we of course have the dominant order inner layer solution and the outer solution on both sides of Γ.

We now proceed to the next order to obtain u_1, U_1, and v_1. For u_1 we have

$$f_u(u_0, x)u_1 = \partial_t u_0 = 0,$$

so

$$\text{(36)}\quad u_1(x) \equiv 0.$$

For U_1 we have, from (31), the problem

(37)

$$\begin{aligned} \partial_{\rho\rho} U_1 + v_0\partial_\rho U_1 + f_u(U_0(\rho, s, t), X)U_1 = (-v_1 - \kappa(s, t))\partial_\rho\psi(\rho, X) \\ + \partial_t\psi(\rho, X(s, t)) + \partial_t s_0\partial_s\psi(\rho, X(s, t)) \\ - \rho\nabla_x f(U_0(\rho, X), X)\cdot e_r, \end{aligned}$$

where in the last term, $\nabla_x f \cdot e_r = \partial_r f$ represents the directional derivative of f with respect to x in the direction of e_r.

The right side of (37) can be simplified, as we now show.

In (35), U_0 is given in terms of the function $\psi(x - vt, x_0)$, defined following (26), with the argument $x - vt$ replaced by ρ and x_0 by $X(s, t)$. Since ψ is defined for any spatial point x_0 occurring as the second argument, we may write

$$U_0(\rho, s, t) = \psi(\rho, x)\,|_{x=X(s,t)}.$$

More generally, consider any smooth function $F(\rho, x, t)$, in which the three arguments are considered to be independent. We examine the effect of the differential operator

$$\hat{D} \equiv \partial_t^{(s)} + (\partial_t s)\partial_s$$

acting on the function $F(\rho, X(s, t), t)$, where as before $X(s, t)$ is the point on Γ described by the arclength coordinate s at time t, and $\partial^{(s)}$ represents differentiation with s held constant. By the chain rule, we have

$$\hat{D}F(\rho, X(s, t), t) = \partial_3 F + \nabla_x F(\rho, X(s, t), t) \cdot [\partial_t X(s, t) + (\partial_t s)\partial_s X(s, t)], \tag{38}$$

where ∂_3 represents differentiation with respect to the third argument of F.

We temporarily denote the dependence of s on x and t by the function $s = S(x, t)$. Thus in particular

$$s = S(X(s, t), t),$$

and the variables s and t are independent in this relation, so we may differentiate with respect to t to obtain

$$0 = \nabla_x S(X, t) \cdot \partial_t X + \partial_t S(X, t). \tag{39}$$

Also note by elementary differential geometry that the vectors $\nabla_x S$ and $\partial_s X$ are (identical) unit vectors tangent to Γ; call that vector T. We therefore have from (39)

$$\begin{aligned} \partial_t X(s, t) + \partial_t s \partial_s X(s, t) &= \partial_t X - (\nabla_x S \cdot \partial_t X)\partial_s X \\ &= \partial_t X - (T \cdot \partial_t X)T = P\partial_t X, \end{aligned} \tag{40}$$

the symbol P denoting projection onto the vector e_r. From (38), (40), and the fact that the norm of the vector $P\partial_t X$ is equal to the normal velocity v, we obtain

$$\hat{D}F = \partial_3 F + \nabla_x F \cdot P\partial_t X = \partial_3 F + v\partial_r F. \tag{41}$$

We now apply (41) to (37) to obtain

$$\begin{aligned} &\partial_{\rho\rho} U_1 + v_0 \partial_\rho U_1 + f_u(U_0(\rho, s, t), X)U_1 \\ &\quad = (-v_1 - \kappa(s, t))\partial_\rho \psi(\rho, X) + v_0 \partial_r \psi(\rho, x)|_{x=X(s,t)} - \rho\partial_r f(U_0(\rho, X), X). \end{aligned} \tag{42}$$

We write the left side of (42) in the form $\mathscr{L}U_1$, where $\mathscr{L}$ is the linear differential operator indicated. It is easily checked that

$$\mathcal{L}p = 0,$$

where $p(\rho, X) \equiv \partial_\rho \psi(\rho, X)$. Thus p is an eigenfunction of $\mathcal{L}$ in $L_2(-\infty, \infty)$. We know that the eigenvalues of $\mathcal{L}$ are simple (see below). Let p^* be the eigenfunction corresponding to eigenvalue 0 of the adjoint operator $\mathcal{L}^*$. Then any equation $\mathcal{L}q = f(\rho)$, $f \in L_2$, is solvable in L_2 if and only if f is orthogonal to p^*. Equation (42) is of that form, with $q = U_1$. We therefore require orthogonality to $p^*(\rho)$:

(43) $$(-v_1 - \kappa)A(s, t) + B(s, t) = 0,$$

where

$$A = \int p(\rho)p^*(\rho)\, d\rho,$$

$$B = \int p^*(\rho)[v_0\partial_r\psi - \rho\partial_r f]\,|_X\, d\rho.$$

The simplicity of the eigenvalue ensures that $A \neq 0$, so that (43) may be written

(44) $$v_1(X) = -\kappa(X) + g(X),$$

for every $X \in \Gamma(t)$, where $g(X(s, t)) = B(s, t)/A(s, t)$, and κ is the curvature of Γ_0 at the point X.

Combining (34) and (44), we obtain a more accurate expression for the normal velocity of Γ in the form

(45) $$v_0(X) + \epsilon v_1(X) = V(X) - \epsilon\kappa(X) + \epsilon g(X).$$

Here, of course, V and g are functions of X that can be determined a priori, but κ depends on Γ itself and is determined as part of the solution. The right side of (45) shows two correction terms to the usual velocity V of a planar front in a homogeneous medium. The first results from the possible curvature of the front, and the second from the inhomogeneous nature of the medium.

It should be noted that this asymptotic result, to lowest order, in the case of one space dimension has recently been rigorously justified by Fife and Hsiao [FH]. The initial formation of the layer was also studied in that paper.

An interesting recent study of the dynamics of interfaces for gradient systems in a homogeneous medium was made by Rubenstein, Sternberg, and Keller [RSK]. They treated systems of the form (24) but with f replaced by the gradient $V_u(u)$ without x dependence.

We return to show that 0 is a simple eigenvalue of $\mathcal{L}$. If q is another eigenvector with that eigenvalue, then the Wronskian $W = pq' - qp'$ satisfies

$$W'' + v_0 W' = 0.$$

But $W \to 0$ at $\pm\infty$, so $W \equiv 0$; hence p and q are linearly dependent.

3. Example: Phase field models.

These models typically assume the form

$$\partial_t u + \lambda \partial_t \phi = \Delta u, \tag{46}$$

$$\alpha \epsilon^2 \partial_t \phi = \epsilon^2 \Delta \phi + f(\phi, u; \epsilon). \tag{47}$$

Here u represents the temperature distribution in a material, and ϕ is an order parameter such that the material is in the solid phase when ϕ is near -1, and in the liquid phase when it is near $+1$. The parameters ϵ, λ, and α (ϵ is small) are a dimensionless interaction length, latent heat, and relaxation time, respectively. For the physical background as well as many other mathematical results, see [Ca], the references therein, and the other papers by Caginalp et al.

The function f is assumed to be bistable in the variable ϕ (see the definition of the h's at the beginning of §2), symmetric (merely for simplicity) when $u = 0$, and to vanish at ± 1 when $u = 0$:

$$\begin{gathered} f(\phi, 0, \epsilon) = -f(-\phi, 0, \epsilon), \quad f(\pm 1, 0, \epsilon) \equiv 0, \\ \int_{h_-(u,\epsilon)}^{h_+(u,\epsilon)} f(\phi, u, \epsilon)\, d\phi \neq 0 \quad \text{if } u \neq 0. \end{gathered} \tag{48}$$

The treatment below follows [CF88]. The interface Γ in this case will be a phase boundary separating the liquid region $\mathscr{D}_+$ from the solid region $\mathscr{D}_-$. As usual, capital letters U and Φ will denote the inner variables corresponding to the outer variables u and ϕ. Near the interface, ϕ will undergo a discontinuity. We specifically define $\Gamma (r = 0)$ to be the location where $\Phi = 0$.

Outer expansion. Setting (2) into (46) and (47) (where u really stands for (u, ϕ)) and equating coefficients of corresponding powers of ϵ, we obtain a sequence of outer problems. They assume the following form when f does not depend on ϵ:

$O(1)$:

$$\partial_t u_0 + \lambda \partial_t \phi_0 = \Delta u_0 \quad (r \neq 0), \tag{49}$$

$$f(\phi_0, u_0) = 0; \tag{50}$$

$O(\epsilon)$:

$$\partial_t u_1 + \lambda \partial_t \phi_1 = \Delta u_1, \tag{51}$$

$$f_\phi(\phi_0, u_0)\phi_1 + f_u(\phi_0, u_0) u_1 = 0; \tag{52}$$

etc.

It will be assumed that (50) can be solved for $\phi_0 = h_+(u_0)$ in a neighborhood of $+1$, for u_0 in the anticipated temperature range in the liquid, and that $f_\phi(\phi_0, u_0) < 0$. In particular, this is the case if $f(\phi, u) = k(\phi - \phi^3) + u$ and $|u_0|$

is not too large. This is then substituted into (49) to obtain an equation for u_0 alone, of the form

$$\partial_t(u_0 + \lambda h_+(u_0)) = \Delta u_0, \tag{53}$$

which should hold on the liquid side of the layer ($r > 0$). This gives a nonlinear heat conduction law in the material, but a linear one can be readily obtained by an elementary modification of the model.

Similarly, it is assumed that (50) can be solved for $\phi_0 = h_-(u_0)$ in a neighborhood of -1; this will generate an equation for u_0 valid for $r < 0$. The higher order equations (51), (52), etc., likewise reduce to equations for u_k alone. To complete the solution, interface conditions at $r = 0$ will be derived later, and boundary and initial conditions may be imposed.

Inner expansion. In terms of ρ, s, and t, (46) and (47) become

$$\begin{aligned}&\partial_{\rho\rho}U + \epsilon(-r_t\partial_\rho U - \lambda r_t \partial_\rho \Phi + \Delta r \partial_\rho U)\\ &\quad - \epsilon^2(\partial_t U + \partial_s U s_t + \lambda\partial_t\Phi + \lambda\partial_s\Phi s_t - (\partial_{ss}U\,|\nabla s|^2 + \partial_s U \Delta s)) = 0,\end{aligned} \tag{54}$$

$$\begin{aligned}&\partial_{\rho\rho}\Phi + f(\Phi, U) - \epsilon\alpha r_t \partial_\rho\Phi + \epsilon\partial_\rho\Phi\Delta r\\ &\quad + \epsilon^2(\partial_{ss}\Phi\,|\nabla s|^2 + \partial_s\Phi\Delta s - \alpha\partial_t\Phi - \alpha\partial_s\Phi s_t) = 0.\end{aligned} \tag{55}$$

We will now proceed to examine the various orders of approximation of (54) and (55) obtained by substituting the expansions analogous to (4) therein.

$O(1)$:

$$\partial_{\rho\rho}U_0 = 0, \tag{56}$$

$$\partial_{\rho\rho}\Phi_0 + f(\Phi_0, U_0) = 0. \tag{57}$$

We want bounded solutions, so from (56), $U_0 = \text{const}$, and the existence of a bounded solution of (57) with distinct limits at $\pm\infty$ requires $U_0 \equiv 0$. To see this, observe that (26) has a traveling wave solution with zero velocity if and only if $\int_{h_-}^{h_+} f(\psi)\,d\psi = 0$; by (48) this is true for the present function f only if $U_0 = 0$. From this and matching condition (5a) we obtain

$$U_0 \equiv u_0\,|_{\Gamma\pm} = 0 \tag{58}$$

and

$$\Phi_0 = \Phi_0(\rho) \equiv \psi(\rho), \tag{59}$$

$\psi(\rho)$ being the unique solution of

$$\psi'' + f(\psi, 0) = 0, \quad \psi(\pm\infty) = \pm 1, \quad \psi(0) = 0. \tag{60}$$

Now (5a) applied to ϕ yields

$$\phi_0\,|_{\Gamma\pm} = \pm 1,$$

which is also evident from (50), (58), and (48).

$O(\epsilon)$:

$$\partial_{\rho\rho} U_1 = r_{0t}\partial_{\rho\rho} U_0 + \lambda\partial_t r_0 \partial_\rho \Phi_0 - \Delta r_0 \partial_\rho U_0 = \lambda r_{0t} \psi'(\rho). \tag{61}$$

The latter equation is a consequence of (58) and (59). Integrating, we obtain

$$\partial_\rho U_1 = \lambda r_{t0} \psi(\rho) + c_1(s, t). \tag{62}$$

But (5b) applies, and we let $\rho \to \pm\infty$ in (62) to obtain

$$\partial_r u_0 \,|_{\Gamma\pm} = \pm\lambda r_{t0} + c_1(s, t). \tag{63}$$

We will replace r_{t0} in (63) and subsequent equations by $-v_0$, and similarly for r_1 and v_1.

Subtracting this equation with the $-$ sign from the same equation with the $+$ sign yields

$$[\partial_r u_0]_\Gamma = -2\lambda v_0 \tag{64}$$

(brackets denote the jump in the value of the function across Γ), which is the Stefan condition for the lowest order approximation. If we are interested in solving an initial boundary value problem, then with appropriate initial conditions, (53), (58), and (64) may be solved to give a unique value for $u_0(x, t)$, $r_0(x, t)$, and $\phi_0(x, t)$.

Once u_0 and r_0 are found this way, they can be used in (63) to determine $c_1(s, t)$. We now proceed toward the determination of U_1. Integrate (62) to obtain

$$U_1(\rho, s, t) = -\lambda v_0 \Psi(\rho) + c_1(s, t)\rho + c_2(s, t), \tag{65}$$

where

$$\Psi(\rho) = \int_0^\rho \psi(z')\, dz'.$$

Thus we obtain the expression

$$\begin{aligned} U_1(\rho, s, t) = \lambda v_0 \int_0^\rho (\operatorname{sgn} z' - \psi(z'))\, dz' \\ - \lambda v_0 |\rho| + c_1(s, t)\rho + c_2(s, t). \end{aligned} \tag{66}$$

Now let $\rho \to \infty$ and use (5b) to obtain (63) again and also

$$u_1 \,|_{\Gamma\pm} = c_2(s, t) + \lambda v_0 \int_0^{\pm\infty} (\operatorname{sgn} z - \psi(z))\, dz. \tag{67}$$

The final determination of u_1 and c_2 (hence U_1) will be done later.

The $O(\epsilon)$ terms in (55) are

$$\partial_{\rho\rho}\Phi_1 + f_\phi(\Phi_0, U_0)\Phi_1 + f_u U_1 = -\alpha v_0 \psi'(\rho) - \kappa_0 \psi'(\rho), \tag{68}$$

where we have used the expression (6) for κ. Let $\Lambda \equiv \partial_{\rho\rho} + f_\phi(\psi(\rho)k, 0)$; then this becomes

$$\Lambda\Phi_1 = -U_1 - \alpha v_0 \psi'(\rho) - \kappa_0 \psi'(\rho). \tag{69}$$

As an operator on $L_2(-\infty, +\infty)$, Λ has an eigenvalue at the origin, with eigenfunction $\psi'(\rho)$. Moreover, it will be simple, so that the solvability condition for $\Lambda\Phi = g \in L_2$ is orthogonality of g to this same eigenfunction, which we denote by ψ' for short. We know from (5a), however, to expect that Φ_1 may be unbounded (growing linearly in z at $\pm\infty$). Nevertheless the solvability condition for (69) remains the same:

$$\int U_1 \psi' \, d\rho + (\alpha v_0 + \kappa_0) \int (\psi')^2 \, d\rho = 0. \tag{70}$$

Now the substitution of (65) into (70) gives the function $c_2(s, t)$ uniquely, hence U_1, since everything else in (66) is known. Specifically, we have

$$c_2 = v_0 \left[\frac{\lambda}{2} \int_{-\infty}^{\infty} \Psi(z) \psi'(z) \, dz - \frac{1}{2} \alpha A \right] - \frac{1}{2} \kappa_0 A, \tag{71}$$

where $A \equiv \int_{-\infty}^{\infty} (\psi')^2 \, dz$. From this and (67), we calculate that

$$u_1 \big|_{\Gamma\pm} = v_0[\lambda B - \tfrac{1}{2}\alpha A] - \tfrac{1}{2} A \kappa_0, \tag{72}$$

where

$$B = \int_0^{\infty} (1 - \psi(z))^2 \, dz. \tag{73}$$

Formula (72) is the dynamic analogue of the Gibbs–Thompson relation. The physical significance of the various terms can be understood [CF88]. For now we proceed with the asymptotic development.

The right side of (72) is a known function of s and t; v_0 and κ_0 have been determined at the stage of the zeroth-order Stefan problem. It provides the needed boundary conditions at the interface to be used with (51), (52), and the imposed boundary and initial conditions in the unique determination of $u_1(x, t)$ on both sides of the interface. If the imposed initial and boundary conditions do not depend on ϵ, then those conditions for u_1 are homogeneous.

Since U_1 is now known, the function Φ_1 can be found uniquely from (69), from the condition that it vanishes when $\rho = 0$, and from the required behavior at $\pm\infty$. The latter, obtained from (5b) applied to Φ_1, turns out to be the same as what we obtain directly from (69), letting $\rho \to \pm\infty$, $\partial_{\rho\rho}\Phi \to 0$, and using the known asymptotic properties of U_1. The orthogonality condition has been imposed, and we obtain a unique solution Φ_1.

This provides the functions u, ϕ, U, Φ to orders 0 and 1, and r to order 0. The process, of course, can be continued.

In summary, to dominant order the motion of the phase interface here is governed by a Stefan problem. The next order approximation, however, brings in effects relating the temperature at the interface to its curvature and normal velocity. Physically, such effects arise from surface tension, which is therefore automatically included in the phase field model (46a), (47).

4. Example: Viscous shocks.

Consider a system of conservation laws in two space dimensions with small dissipation added:

$$\partial_t u + \partial_{x_1} f_1(u) + \partial_{x_2} f_2(u) = \epsilon D \Delta u, \tag{74}$$

where $u(x, t) \in R^n$, f_1 and f_2 are functions from R^n into R^n, D is a (dissipation) matrix, and ϵ as always is a small positive number.

The lowest order outer solution $u_0(x, t)$ satisfies the conservation laws without dissipation, and of course we are interested in solutions with a single shock along a curve $\Gamma(t)$. Not much of a rigorous nature is known about curved shocks. Assuming one exists, our aim here is merely to examine the effect of curvature on the shock relations. Work on the rigorous justification of layer analysis for similar problems in one space dimension can be found in [Ga] and [Hab].

The solution near the shock is represented by the inner function $U(\rho, s, t; \epsilon)$. To obtain the inner problems we transform to the coordinates ρ, s, t. To the lowest two orders in ϵ, (74) then becomes

$$D\partial_{\rho\rho} U + v\partial_\rho U - \partial_\rho f^{(r)}(U) = \epsilon(-D\kappa\partial_\rho U + \partial_s f^{(s)} + \partial_t U) \tag{75}$$

where $f^{(r)} = v_{rx_1} f_1 + v_{rx_2} f_2$, and v_{rx_1} is the x_1 component of the unit vector in the direction of increasing r, i.e., in the direction normal to Γ (a similar definition for $f^{(s)}$). The lowest order term U_0 satisfies

$$D\partial_{\rho\rho} U_0 + v_0 \partial_\rho U_0 - \partial_\rho f^{(r)}(U_0) = 0. \tag{76}$$

This equation is to be solved under the boundary conditions $U_0(\pm\infty, s, t) = u_0(\Gamma\pm, s, t)$. If it has a solution, then integration of (76) with respect to ρ from $-\infty$ to ∞ gives the usual shock condition

$$v_0[u_0]_\Gamma = [f^{(r)}(u_0)]_\Gamma. \tag{77}$$

Here the brackets signify the jump across Γ of the discontinuous function inside the brackets. This does not provide much new information. It simply says that if a single shock solution of the system of conservation laws exists, as well as a solution with small viscosity, and if (76) has a solution with the appropriate boundary conditions, then that viscous shock solution is compatible with the original shock conditions satisfied by the solution with single shock.

More information is provided by the higher order terms; it is there that the character of the dissipation matrix exerts an influence. The next order inner equation is

$$
\begin{aligned}
(78)\qquad & D\partial_{\rho\rho}U_1 + v_0\partial_\rho U_1 - \partial_\rho(df^{(r)}(U_0)U_1) \\
&\qquad = -v_1\partial_\rho U_0 + \partial_s f^{(s)}(U_0) + \partial_t U_0 - \kappa D\partial_\rho U_0.
\end{aligned}
$$

Let the left side of this equation be denoted by $\mathscr{L}U_1$. Then $\mathscr{L}$ has a nontrivial nullfunction $p \equiv \partial_\rho U_0$ (seen by differentiating (76)), and its adjoint $\mathscr{L}^*$ also has a nontrivial nullfunction. Let us assume that 0 is a simple eigenvalue, and let p^* be the nullfunction of $\mathscr{L}^*$. Then solvability of (78) is dependent on the satisfaction of the orthogonality condition

$$\int_{-\infty}^{\infty} [-(v_1 + \kappa D)p + \partial_s f^{(s)}(U_0) + \partial_t U_0] \cdot p^*(\rho)\, d\rho = 0,$$

i.e.,

$$(79)\qquad (v_1 + \kappa D)\int pp^*\, d\rho = \int p^*(\rho) \cdot [\partial_s f^{(s)} + \partial_t U_0]\, d\rho.$$

This formula shows that the velocity correction v_1 depends linearly on the curvature κ of the shock, and also depends on the transverse flux $f^{(s)}$.

CHAPTER 2

Flame Theory

Interfaces are often found in spatially distributed combustion phenomena. They form an integral part of many, if not most, mathematical studies in this field. In §1 we review the well-known arguments of activation energy asymptotics for a simple one-dimensional premixed flame arising from a one-step exothermic chemical reaction. We do this analysis within the context of the general formalism of §1 in Chap. 1. The interface separates two zones in the flame profile: the preheat zone, where the temperature is low enough so that no chemical reaction has yet occurred; and the burned zone, where the gas has attained its final state. The interface between them is the combustion zone, and it is thin because the reaction rate depends very strongly on temperature (this is the high activation energy assumption).

In higher dimensions, the stability analysis of the planar interface examined in §1 is a crucial consideration, because in most practical situations it is unstable. This analysis, within the framework of the thermodiffusive model, and again for a one-step reaction, is given in §2. The analysis is well known, but the treatment given here is a much expanded version designed to bring out the details of various parts of the argument that are often elusive when first read. read.

The production and diffusion of heat is the chain effect that is typically mainly responsible for the flame's propagation in space. However, in §3 a second contributing factor to this propagation is brought out. It is the production of radicals in the combustion zone, which diffuse upstream and induce further reactions there.

Flame theory is complicated in a very real way by the presence of many, rather than one or two, chemical reactions. This complication within the flame layer is addressed in §4, where one approach is described for the systematic construction of simpler chemical models to replace more realistic but complex ones.

1. Activation energy asymptotics for a one-step reaction.

The flame profiles of interest here satisfy the same differential equations as traveling waves for a thermodiffusive model, to be given below. Although that model has the deficiency of not accounting for density variation of the gas across the flame, its traveling wave solutions are equivalent to those of a more exact model in which density variation is accounted for, but the pressure is taken to be constant (a good approximation for slow deflagrations). In the next section, a stability analysis will be reviewed within the context of the thermodiffusive model. For convenience we will operate within the formalism of that simpler model even in the present section where we deal with steady deflagrations.

The chemical reaction supporting the flame is the one-step exothermic reaction

$$A \rightarrow \text{Products}.$$

As the temperature variable we choose $\theta \equiv (T - T_-)/(T_+ - T_-)$, T_- and T_+ being the unburned (fresh) and burned temperatures (Kelvin), respectively.

For the concentration variable we use Y, the mass fraction of reactant A in the gas divided by the mass fraction in the fresh mixture. The thermodiffusive model in this case is the pair of equations

$$\partial_\tau \theta = \Delta\theta + \epsilon^{-1}\omega(\theta, Y, \epsilon), \tag{1}$$

$$\partial_\tau Y = L^{-1}\Delta Y - \epsilon^{-1}\omega(\theta, Y, \epsilon), \tag{2}$$

where the Lewis number L represents the ratio between thermal and material diffusivities, ω is a scaled reaction rate, and ϵ is the reciprocal of the "Zel'dovich number," $\epsilon^{-1} = E(T_+ - T_-)/RT_+^2$, E being the activation energy and R the gas constant. It will be assumed that

$$0 < \epsilon \ll 1.$$

In (1) and (2) the space and time scales can be chosen in such a way that the scaled reaction rate ω is

$$\omega = \epsilon^{-1}Y(\exp(\epsilon^{-1}(\theta - 1))(1 + 2\gamma\epsilon^{-1}(\theta - 1)^2)), \tag{3}$$

where $\gamma = (T_+ - T_-)/T_+$. The derivation of (3) from an Arrhenius law will be given at the end of this section.

This model is standard (see [BL82], [BL83], for instance). The asymptotics given in this section were put on a rigorous foundation by Berestycki, Nicolaenko, and Scheurer [BNS85].

Let the physical space variables be denoted by $\bar{x}$ and z, and suppose the flame moves in the direction of the negative $\bar{x}$ axis. In this section we suppose it is planar, so use the traveling wave variable $x = \bar{x} + v\tau$, where $-v$ is the velocity. In one space dimension the Laplacian in (1), (2) is simply the second derivative with respect to $\bar{x}$. The traveling wave equations for (1), (2) are the following:

$$\partial_{xx}\theta - v\partial_x\theta + \epsilon^{-1}\omega = 0, \tag{4a}$$

$$L^{-1}\partial_{xx}Y - v\partial_x Y - \epsilon^{-1}\omega = 0. \tag{4b}$$

The boundary conditions are

$$\theta(-\infty)=0, \quad \theta(\infty)=1, \quad Y(-\infty)=1, \quad Y(\infty)=0. \tag{5}$$

For any fixed $\theta<1$, the exponential decay of (3) makes ω negligible to all finite orders in ϵ. For θ within $O(\epsilon)$ of 1, on the other hand, $\epsilon^{-1}\omega$ is large of order ϵ^{-1}. This has the effect of inducing a large curvature in the flame profile near $\theta=1$; in that region, the profile is described most naturally in terms of stretched coordinates. That will be the inner layer, which we specify to be located at $x=0$. The outer problems for $x<0$, therefore, are constructed by expanding

$$\theta=\Sigma\epsilon^n\theta_n, \quad Y=\text{(similar)}, \quad v=\text{(similar)} \tag{6}$$

and substituting into (4) with ω set equal to 0.

The stretched spatial coordinate will be $\xi=\epsilon^{-1}x$, and the inner temperature variable could be taken as $\Theta(\xi)$, in accordance with the formalism in Chap. 1, §1, so that formally

$$\Theta(\xi;\epsilon)\simeq\theta(\epsilon x;\epsilon)=\Theta_0(\xi)+\epsilon\Theta_1(\xi)+\epsilon^2\Theta_2(\xi)+\cdots. \tag{7}$$

However, we will take $\Theta_0\equiv 1$, and for convenience and in accordance with more accepted practice, we use the inner variable

$$t(\xi;\epsilon)\equiv\epsilon^{-1}(\Theta-1)\equiv\Theta_1+\epsilon\Theta_2+\cdots\equiv t_0(\xi)+\epsilon t_1(\xi)+\cdots, \tag{8a}$$

so that in fact we are identifying $t_0=\Theta_1$, etc. In the same way, we define the inner concentration variable

$$y(\xi;\epsilon)\equiv\epsilon^{-1}Y=y_0(\xi)+\epsilon y_1(\xi)+\cdots. \tag{8b}$$

Then the inner equations become

$$\partial_{\xi\xi}t+ye^t+\epsilon[2\gamma t^2ye^t-v\partial_\xi t]+O(\epsilon^2)=0, \tag{9a}$$

$$L^{-1}\partial_{\xi\xi}y-ye^t+\epsilon[-2\gamma t^2ye^t-v\partial_\xi y]+O(\epsilon^2)=0. \tag{9b}$$

To lowest order in ϵ, we have the outer problem on the left (for $x<0$)

$$\partial_{xx}\theta_0-v_0\partial_x\theta_0=0, \tag{10a}$$

$$L^{-1}\partial_{xx}Y_0-v_0\partial_xY_0=0, \tag{10b}$$

subject to the boundary conditions at $-\infty$ given by (5), and the inner problem

$$\partial_{\xi\xi}t_0+y_0e^{t_0}=0, \tag{11a}$$

$$L^{-1}\partial_{\xi\xi}y_0-y_0e^{t_0}=0. \tag{11b}$$

For boundary conditions, we will attempt to have $t_0(\infty)=y_0(\infty)=0$, which, in view of the definition of the inner variables, is tantamount to requiring that the boundary conditions at $+\infty$ given by (5) be satisfied already in the inner layer.

The solution of (10) is then

$$\theta_0 = a_0 e^{v_0 x},$$

$$Y_0 = 1 - b_0 e^{L v_0 x},$$

for some constants a_0 and b_0. By the matching condition (5a) in Chap. 1, §1, and the fact that we have chosen $\Theta_0 \equiv 1$ and the dominant inner concentration to be zero, we have $a_0 = b_0 = 1$, so

$$\theta_0 = e^{v_0 x}, \tag{12a}$$

$$Y_0 = 1 - e^{L v_0 x}. \tag{12b}$$

The solution of (11) is obtained as follows. Adding the two, we have that the second derivative of $(t_0 + L^{-1} y_0)$ vanishes identically, and since it is bounded it must equal a constant. Evaluating the constant at $+\infty$, in view of the stated boundary conditions, gives us zero. Therefore

$$y_0 \equiv -L t_0, \tag{13}$$

and from (11a),

$$\partial_{\xi\xi} t_0 = L t_0 e^{t_0}. \tag{14}$$

This is solved by multiplying by $\partial_\xi t_0$ and integrating. Specifically, we get

$$\partial_\xi t_0(-\infty) = \sqrt{2L}. \tag{15}$$

According to the matching condition (5b) in Chap. 1, §1, this is to be identified with $\partial_x \theta_0(0-)$, which by (12a) is v_0. The same matching applied to y_0 gives the same result. We therefore have the dominant order velocity

$$v_0 = \sqrt{2L}. \tag{16}$$

Indeed it is true that the inner variables approach the correct boundary condition at $+\infty$. Therefore we could say, according to our preference, that either (i) the inner region extends to ∞, since there is no outer region to the right, or (ii) the outer region to the right of 0 has, as outer variables, $\theta_0 \equiv 1$, $Y_0 = 0$.

The next order correction to the profile and to v will be obtained in the next section.

A word or two should be said about the "cold boundary difficulty," which is hardly a difficulty at all. The term refers to the fact that our model (4), (5) has no exact solution on the infinite interval, since ω is bounded away from zero for positive Y (i.e., near $x = -\infty$), although it is small to exponential order in ϵ. What we have done is ignore that fact and constructed an "asymptotic" solution anyway. To state the case more accurately and fairly, we have replaced the original model (4), (5) (which is actually self-contradictory because we have abstracted the physical length of the system to be infinite) by another model which is less accurate in some respects, but on the other hand does not suffer from the stated inconsistency. (It is not necessarily logical to expect models built from unrealistic assumptions, like "the domain is infinite," to yield consistent mathematics.) This replacement was done with the aid of formal asymptotics, which is a popular tool for "remodeling" purposes. Of course, there are other ways to cure the cold boundary difficulty, but they all involve revising the original inconsistent model.

Finally, the derivation of (3) will be sketched. An acceptable rate expression for a chemical reaction is the Arrhenius law $\hat{\omega}(Y, T) = BYe^{(-E/RT)}$ where B is some constant pre-exponential factor and the other symbols were defined following (2). We expand the exponent around $T = T_+$:

$$\begin{aligned} -\frac{E}{RT} &= -\frac{E}{RT_+} + \frac{E}{RT_+^2}(T - T_+) - \frac{2E}{RT_+^3}(T - T_+)^2 + \cdots \\ &= -\frac{E}{RT_+} + \epsilon^{-1}(\theta - 1) + 2\epsilon^{-1}\gamma(\theta - 1)^2 \cdots . \end{aligned} \tag{17}$$

Now let $t \equiv \epsilon^{-1}(\theta - 1)$. Then

$$e^{-E/RT} \simeq e^{-E/RT_+}e^t(1 + 2\epsilon\gamma t^2 + \cdots). \tag{18}$$

It will turn out that θ will never surpass 1 in the flame, so $t \leq 0$ always. For $|t| \leq \epsilon^{-1/4}$, for example, a good approximation can be made by truncating at two terms in (18). But for $|t| > \epsilon^{-1/4}$ that is also true because the presence of the factor e^t in (18) renders the right side very small, whether or not the series is truncated. Using that truncation, we obtain (3) except for a factor Be^{-E/RT_+}. That factor can be reduced, and was implicitly, to unity by rescaling space and time (more exactly, by rescaling the variable x and the velocity v).

2. Stability analysis of simple flames.

Our object here is to explain the standard linear stability analysis of the simple flames constructed in the last section. Although this somewhat involved calculation, and modifications of it, has appeared a number of times in the literature beginning with Sivashinsky (e.g., [Si77a], [MM], [JC79], [CFN], [BL83]), many details of the argument are usually omitted. The account given here is much longer than previous treatments. We emphasize that this is necessary to clarify various steps in the reasoning, which otherwise may be confusing to the reader unaccustomed to this type of analysis.

Generally there are at least two mechanisms considered that may operate on flames to make them unstable. One is associated with hydrodynamic effects, in which thermal expansion is an all-important consideration. These effects are disregarded in the thermodiffusive model. That model, however, has its own quite different mechanisms which produce instabilities of a cellular nature. Although admittedly these last instabilities are not the whole story, they nevertheless do account well for phenomena observed in the laboratory. They are modeled by the system investigated in this section.

2A. Formulation. The starting point is the system (1), (2), (3). In the last section a strict traveling wave solution was constructed asymptotically to lowest order in ϵ. It did not, of course, depend on the other space coordinate z, nor on τ except through the traveling wave coordinate x. We now allow separate dependence on z and τ, but at the same time utilize the same traveling wave coordinate $x = \bar{x} + v\tau$, so that θ and Y will be functions of x, z, and τ. However, we make one further modification of the coordinate system. Whereas the steady wave had a straight inner layer at $x = 0$, that layer may now be curved, its position denoted by $x = \delta\phi(z, \tau)$. Here δ is a small parameter measuring the size of the nonplanar deviation.

We must be somewhat more precise about the meaning of ϕ since the layer's position is not completely defined. As before, it will turn out that the inner variable y will approach its limiting value of zero already in the layer: $\lim_{\xi\to\infty} y(\xi) = 0$. This, in fact, will occur to all orders. The *outer* solution $Y(x, z, \tau; \epsilon)$ (on the left) will vanish at some value $x = x_0(z, \tau; \epsilon)$. We simply choose $\delta\phi$ to be exactly that value. The shape ϕ of the deviation may be expected to depend on its amplitude δ and on ϵ, so we will think of $\phi = \phi(z, \tau; \delta, \epsilon)$. We now center a new coordinate system about the curved interface by defining the new variable

$$x' \equiv x - \delta\phi.$$

In terms of the variables x', z, and τ, the system (1), (2), (3) now takes the following form, where we have dropped the primes and omitted all terms quadratic or higher order in δ:

$$\partial_\tau\theta - \delta\phi_\tau\partial_x\theta + v\partial_x\theta = \Delta_{xz}\theta - \delta[2\phi_z\partial_{xz}\theta + \phi_{zz}\partial_x\theta] + \epsilon^{-1}\omega, \tag{19}$$

$$\partial_\tau Y - \delta\phi_\tau\partial_x Y + v\partial_x Y = L^{-1}\Delta_{xz}Y - L^{-1}\delta[2\phi_z\partial_{xz}Y + \phi_{zz}\partial_x Y] - \epsilon^{-1}\omega. \tag{20}$$

We conceive of solving a layered initial value problem for (19) and (20), i.e., θ, Y, and ϕ are all prescribed initially and we want to find how they evolve in time. However, the initial data are supposed to be a perturbation of the steady solution found earlier, so we require that the solution be precisely that solution if $\delta = 0$.

According to (3), we may write

$$\omega = ye^t(1 + 2\epsilon\gamma t^2), \tag{21a}$$

where, as in (8),

$$y = \epsilon^{-1}Y, \qquad t = \epsilon^{-1}(\theta - 1). \tag{21b}$$

The important stability changes occur for L in an $O(\epsilon)$ neighborhood of 1, so we define a scaled Lewis number l by setting

$$L = 1 + \epsilon l. \tag{22}$$

There are now two small parameters, ϵ and δ, appearing explicitly and implicitly in (19)–(21). We will consider them to be independent parameters, and formally expand the solution in a power series in both of them.

The outer functions θ and Y, the deviation ϕ, and the velocity v thus will be expanded as

$$\theta \simeq \Sigma \delta^i \epsilon^j \theta_j^i(x, z, \tau), \quad Y \simeq (\text{similar}), \quad \phi \simeq (\text{similar}), \quad v \simeq \Sigma \epsilon^j v_j.$$

The reason v is not taken to depend on δ is that any δ-dependence of the velocity or position of the front can be incorporated into the function ϕ. Of course this also could be done for the ϵ-dependence, but will not be because of our desire stated above that when $\delta = 0$, we obtain the steady solution of the previous section, which *does* depend on ϵ. Therefore, $v(\epsilon)$ will be the velocity of the unperturbed flame. In the same way, when $\delta = 0$, the above expansions are supposed to yield the ϵ-dependent steady flame. Similar expansions will hold for the inner variables t and y, defined in (21b).

As usual, there will be outer and inner problems for each of the double-indexed variables. We will denote the outer and inner problems of order $\delta^i \epsilon^j$ by O_j^i and I_j^i, respectively.

2B. Outer problems. As in §1, the reaction term ω is omitted in the outer region. In fact, we shortly will see that $t \to -\infty$ linearly to dominant order as the inner variable $\xi = \epsilon^{-1} x \to -\infty$, and its effect is exponentially small in the outer region to the left. Similarly, to the right it will turn out that $y \to 0$ exponentially as $\xi \to \infty$ (to all orders), so again from (21a), ω will be zero to all finite orders in the outer zone to the right.

$\boxed{O_0^0}$ Using the differential operator defined by $\mathscr{L}\theta \equiv \partial_\tau \theta + v_0 \partial_x \theta - \Delta_{xz} \theta$, we have

$$\mathscr{L}\theta_0^0 = 0, \qquad x \neq 0, \tag{23a}$$

$$\mathscr{L}Y_0^0 = 0, \qquad x \neq 0, \tag{23b}$$

$$\theta_0^0(-\infty) = 0, \quad \theta_0^0(\infty) = 1, \quad Y_0^0(-\infty) = 1, \quad Y_0^0(\infty) = 0. \tag{23c}$$

In the following problems, it is understood that the equations hold for $x \neq 0$, and zero boundary conditions are satisfied at $\pm\infty$.

$\boxed{O_1^0}$

$$\mathscr{L}\theta_1^0 = -v_1 \partial_x \theta_0^0, \tag{24a}$$

$$\mathscr{L}Y_1^0 = -l \Delta_{xz} Y_0^0 - v_1 \partial_x Y_0^0. \tag{24b}$$

$\boxed{O_0^1}$

Here we anticipate that θ_0^0, θ_1^0, Y_0^0, and Y_1^0 will depend only on x, and so we will omit terms involving their derivatives with respect to z or τ.

$$\mathscr{L}\theta_0^1 = \partial_\tau \phi_0^0 \partial_x \theta_0^0 - \partial_{zz} \phi_0^0 \partial_x \theta_0^0, \tag{25a}$$

$$\mathscr{L}Y_0^1 = \partial_\tau \phi_0^0 \partial_x Y_0^0 - \partial_{zz} \phi_0^0 \partial_x Y_0^0, \tag{25b}$$

$\boxed{O_1^1}$

(26a) $$\mathcal{L}\theta_1^1 = (\partial_\tau - \partial_{zz})\phi_0^0\partial_x\theta_1^0 + (\partial_\tau - \partial_{zz})\phi_1^0\partial_x\theta_0^0 - v_1\partial_x\theta_0^1,$$

(26b) $$\begin{aligned}\mathcal{L}Y_1^1 = (\partial_\tau - \partial_{zz})\phi_0^0\partial_x Y_1^0 + (\partial_\tau - \partial_{zz})\phi_1^0\partial_x Y_0^0 - v_1\partial_x Y_0^1 \\ + l[-\Delta_{xz} Y_0^1 + \partial_{zz}\phi_0^0\partial_x Y_0^0].\end{aligned}$$

2C. Inner problems. We change dependent variables according to (21b), and the independent variable by

(27) $$\xi = \epsilon^{-1}x.$$

The basic equations (19), (20), with (21a) taken into account, are thereby transformed to

(28a) $$\begin{aligned}\partial_{\xi\xi}t + ye^t + \epsilon[2\gamma t^2 ye^t - v\partial_\xi t] \\ + \delta\epsilon[\phi_\tau\partial_\xi t - 2\phi_z\partial_{\xi z}t - \phi_{zz}\partial_\xi t] + O(\epsilon^2) = 0,\end{aligned}$$

(28b) $$\begin{aligned}\partial_{\xi\xi} y - ye^t + \epsilon[-l\partial_\xi^2 y - 2\gamma t^2 ye^t - v\partial_\xi y] \\ + \delta\epsilon[\phi_\tau\partial_\xi y - 2\phi_z\partial_{\xi z} y - \phi_{zz}\partial_\xi y] + O(\epsilon^2) = 0.\end{aligned}$$

$\boxed{I_0^0}$

(29a) $$\partial_{\xi\xi}t_0^0 + y_0^0 e^{t_0^0} = 0,$$

(29b) $$\partial_{\xi\xi} y_0^0 - y_0^0 e^{t_0^0} = 0.$$

$\boxed{I_1^0}$ Now define these differential operators acting on pairs of functions:

$$\mathcal{M}_1(t, y) \equiv \partial_{\xi\xi}t + (ye^t + ye^t t),$$

$$\mathcal{M}_2(t, y) \equiv \partial_{\xi\xi} y - (ye^t + ye^t t);$$

then

(30a) $$\mathcal{M}_1(t_1^0, y_1^0) = v_0\partial_\xi t_0^0 - 2\gamma(t_0^0)^2 y_0^0 e^{t_0^0},$$

(30b) $$\mathcal{M}_2(t_1^0, y_1^0) = v_0\partial_\xi y_0^0 + 2\gamma(t_0^0)^2 y_0^0 e^{t_0^0} + l\partial_{\xi\xi} y_0^0.$$

$\boxed{I_0^1}$

(31a) $$\mathcal{M}_1(t_0^1, y_0^1) = 0,$$

(31b) $$\mathcal{M}_2(t_0^1, y_0^1) = 0.$$

$\boxed{I_1^1}$

Here we anticipate that t_0^0 and y_0^0 depend only on ξ, and omit other derivatives.

(32a) $$\partial_{\xi\xi}t_1^1 + (ye^t)_1^1 + 2\gamma(t^2ye^t)_0^1 - v_0\partial_\xi t_0^1 + (\partial_\tau - \partial_{zz})\phi_0^0\partial_\xi t_0^0 = 0,$$

(32b) $$\partial_{\xi\xi} y_1^1 - l\partial_{\xi\xi} y_0^1 - (ye^t)_1^1 - 2\gamma(t^2ye^t)_0^1 - v_0\partial_\xi y_0^1 + (\partial_\tau - \partial_{zz})\phi_0^0\partial_\xi y_0^0 = 0.$$

(The second term in (32a), for example, is defined as the coefficient of $\delta\epsilon$ in the expansion of the function ye^t.)

2D. Matching relations. We apply (5) in Chap. 1, §1, for each fixed *superscript,* to our functions θ and t, noting the relation (8a) between t and the originally conceived inner variable Θ. We also apply the analogous reasoning to Y and y. The results are (37)–(44), but in preparation for deriving them we develop some a priori relations involving the outer functions. (For convenience, we disregard dependence of the functions on z and τ in the following notation.)

First, recall from (8a) that the dominant order inner variable $\Theta_0^0 \equiv 1$ in the unperturbed case ($\delta = 0$) considered there. We will take that to be true now also, to all orders in δ: $\Theta_0^i = 0$, $i \geq 1$. In fact, if this were not done, we observe from (21a) and the fact that ϵ and δ are independent that the reaction term $\epsilon^{-1}\omega$ in (19) and (20), which takes the form $\epsilon^{-2} Y \exp(\epsilon^{-1}(\Theta - 1))$, would be exponentially large as $\epsilon \to 0$ for positive (albeit small) δ; the inner equations then could not be solved. From the matching relation (5a) in Chap. 1, §1, we therefore obtain

$$\theta_0^j(0\pm) = 0, \quad \text{all } j > 0. \tag{33}$$

When $\delta = 0$, the outer equations for θ (which will be a function of x alone) take the form $\partial_{xx}\theta = v\partial_x\theta$, with $v > 0$. The only bounded solution for positive x is a constant, which must be taken equal to the boundary condition at $+\infty$. Therefore $\theta_i^0(x)$ is 0 (except when $i = 0$, when it is one) for $x > 0$:

$$\theta_i^0(x) \equiv 0, \quad \text{all } i \geq 1, \quad \text{for } x > 0. \tag{34}$$

Similar relations hold for the outer variable Y for different reasons. By our definition of ϕ as the place where the outer variable Y on the left vanishes to all orders, we have

$$Y(0-) = 0 \quad \text{to all orders.} \tag{35}$$

Consider now the inner equation (28b) for y. If we set $\epsilon = 0$ in that equation, clearly any solution that grows in at most a polynomial fashion in ξ would decay exponentially with all its derivatives as $\xi \to \infty$, because the factor e^t in the second term is always positive and bounded away from zero. This exponential decay, moreover, does not change when ϵ is nonzero but small enough. Therefore, a priori, we know that y and its derivatives vanish to all orders at $+\infty$. By matching with the outer solution, we therefore obtain

$$Y(x) \equiv 0 \quad \text{to all orders,} \quad \text{for } x > 0. \tag{36}$$

The remaining specific matching conditions are the following, where the symbol $\simeq$ means that the function on the left has the indicated behavior as $\xi \to \pm\infty$, the sign being the same as the sign in the arguments on the right, and $\partial \equiv \partial_x$. The above a priori relations (33)–(36) are incorporated below.

$$t_0^0 \simeq \partial\theta_0^0(0\pm)\xi + \theta_1^0(0\pm), \tag{37}$$

$$y_0^0 \simeq \partial Y_0^0(0\pm)\xi, \tag{38}$$

(39) $$t_1^0 \simeq \tfrac{1}{2}\xi^2\partial^2\theta_0^0(0\pm) + \xi\partial\theta_1^0(0\pm) + \theta_2^0(0\pm),$$

(40) $$y_1^0 \simeq \tfrac{1}{2}\xi^2\partial^2 Y_0^0(0\pm) + \xi\partial Y_1^0(0\pm),$$

(41) $$t_0^1 \simeq \partial\theta_0^1(0\pm)\xi + \theta_1^1(0\pm),$$

(42) $$y_0^1 \simeq \partial Y_0^1(0\pm)\xi,$$

(43) $$t_1^1 \simeq \tfrac{1}{2}\xi^2\partial^2\theta_0^1(0\pm) + \xi\partial\theta_1^1(0\pm) + \theta_2^1(0\pm),$$

(44) $$y_1^1 \simeq \tfrac{1}{2}\xi^2\partial^2 Y_0^1(0\pm) + \xi\partial Y_1^1(0\pm).$$

Note that (34), (36), and the above imply

(45) $$t_j^0 \simeq 0 \quad (\xi\to\infty), \quad \text{all } j,$$

and as mentioned before,

(46) $$y_j^i \simeq 0 \quad (\xi\to\infty), \quad \text{all } i \text{ and } j.$$

2E. Solutions of the outer and inner problems. In the following, semicolons will serve to separate expressions for the outer functions for $x>0$ (on the right of the semicolon) from those for $x<0$ (on the left). The notation t_-, ∂t_-, $\partial^2 t_-$, for any function $t(\xi)$, denotes the coefficients A, B, C in the asymptotic expression $t \simeq A + B\xi + \frac{1}{2}C\xi^2$ $(\xi\to -\infty)$. This is also true for the plus sign.

O_0^0:

(47a) $$\theta_0^0(x) = e^{v_0 x};\, 1,$$

(47b) $$Y_0^0(x) = 1 - e^{v_0 x};\, 0.$$

I_0^0:

(48) $$y_0^0 = -t_0^0, \quad \partial_{\xi\xi}t_0^0 = t_0^0 e^{t_0^0}, \quad t_0^0(\infty) = 0.$$

Thus multiplying by $\partial_\xi t_0^0$ and integrating from $-\infty$ to ∞, we get

(49) $$\partial t_{0-}^0 = 2^{1/2} = -\partial y_{0-}^0.$$

The solution vanishes exponentially as $\xi\to\infty$, in accordance with (36). From this and also by (35), (48), we have

(50) $$t_{0-}^0 = t_{0+}^0 = y_{0-}^0 = y_{0+}^0 = 0,$$

and in fact (here H is the Heaviside function)

(51) $$t_0^0(\xi) - 2^{1/2}\xi H(-\xi) \simeq 0 \quad \text{to all finite orders as } \xi\to\pm\infty.$$

From (37_-), (49), and (47a) we also have

(52) $$v_0 = 2^{1/2}.$$

O_1^0:

From (24), (47), (52), (34),

(53a) $$\theta_1^0 = (v_1 x e^{\sqrt{2}x} + a_1^0 e^{\sqrt{2}x});\, 0$$

where $a_1^0 \equiv \theta_1^0(0-)$,

(53b) $$Y_1^0 = -(v_1 + 2^{1/2}l)xe^{\sqrt{2}x};\ 0.$$

I_1^0:
Adding (30a) and (30b), we obtain

$$\partial_{\xi\xi}(t_1^0 + y_1^0 + lt_0^0) = 0,$$

and in view of (45), (46),

(54) $$t_1^0 + y_1^0 + lt_0^0 = 0,$$
$$y_1^0 = -t_1^0 - lt_0^0.$$

Defining the operator $\mathcal{M}$ by $\mathcal{M}t \equiv [\partial_{\xi\xi} - e^t(t+1)]t$, we obtain from (54) and (30a)

(55) $$\mathcal{M}t_1^0 = lt_0^0 e^{t_0^0} + 2^{1/2}\partial_\xi t_0^0 + 2\gamma(t_0^0)^3 e^{t_0^0}.$$

Observe also from differentiating (48) and setting

$$p(\xi) \equiv \partial_\xi t_0^0(\xi)$$

that

(56) $$\mathcal{M}p = 0.$$

From this and (55),

(57) $$\partial_\xi(p\partial_\xi t_1^0 - t_1^0\partial_\xi p) = p\cdot(\text{right-hand side of (55)}).$$

From (51), we know that

(58) $$p = 2^{1/2}H(-\xi) + \hat{p}(\xi),$$

where $\hat{p}$ approaches zero exponentially as $\xi \to \pm\infty$. We now use this expression in (57) and integrate the latter from ξ to ∞, noting

$$pt_0^0 e^{t_0^0} = \partial_\xi((t_0^0 - 1)e^{t_0^0}),$$

etc., to obtain

$$-p(\xi)\partial_\xi t_1^0(\xi) + t_1^0(\xi)\partial_\xi p(\xi) = l\int_{t(\xi)}^0 se^s\,ds - 2^{3/2}\xi$$
$$+ 2^{1/2}\int_\xi^\infty (2^{3/2}H(-s)\hat{p}(s) + \hat{p}^2(s))\,ds + 2\gamma\int_{t(\xi)}^0 s^3 e^s\,ds.$$

Below, we set

$$Q' \equiv \int_{-\infty}^\infty (2^{1/2}H(-\xi)\hat{p} + \hat{p}^2)\,d\xi.$$

Now let $\xi \to -\infty$, with use of (49) and the fact that $t_0^0 \to -\infty$ to obtain

(59a) $$\partial^2 t_{1-}^0 = 2,$$

(59b) $$\partial t_{1-}^0 = -Q' - l - 12\gamma \equiv Q,$$

and from (54),

(59c) $$\partial y^0_{1-} = -\partial t^0_{1-} - l2^{1/2}.$$

From (39) and (40) we can identify the expressions on the right of (59) with derivatives calculated from (53). Specifically, from (39) we obtain $\partial^2\theta^0_0(0-) = 2$, which checks with (47a) and (52). Also from (39) and (53a) we have

$$\partial\theta^0_1(0-) = v_1 + a^0_1 2^{1/2} = -Q.$$

Similarly from (40) and (53b) we get

$$\partial Y^0_1(0-) = -(v_1 + 2^{1/2}l) = Q - 2^{1/2}l.$$

The conclusion from these two equations is that

(60) $$a^0_1 = 0, \qquad v_1 = -Q.$$

In all, from (53) and (60) we have

(61) $$\theta^0_1 = -Qxe^{\sqrt{2}x};\ 0, \qquad Y^0_1 = (Q - 2^{1/2}l)xe^{\sqrt{2}x};\ 0.$$

O^1_0:

Note that (25a) is linear in the unknowns θ^1_0 and ϕ^0_0, with coefficients that do not depend on τ or z. The general formal solution of that equation, therefore, could be obtained by taking the Fourier transform in z and the Laplace transform in τ. Our stability analysis will proceed by examining each mode in such a Fourier–Laplace decomposition. These modes for the functions θ^1_0 and ϕ^0_0 will be of the form $e^{ikz+\sigma\tau}$ times functions of x (except that ϕ^0_0 will not have the function of x). The same will be true of the function Y^1_0 and the higher order terms θ^1_1 and Y^1_1. We use the symbol $\phi \equiv \phi^0_0$ for simplicity. Accordingly, we set

(62) $$\phi \equiv e^{ikz+\sigma\tau},$$

(63) $$\theta^1_0 = \phi\tilde{\theta}^1_0(x), \qquad Y^1_0 = \phi\tilde{Y}^1_0(x),$$

and later we will use similar expressions for θ^1_1 and Y^1_1. If we define $\mathcal{N} \equiv (-\partial_{xx} + 2^{1/2}\partial_x + \sigma + k^2)$, then (25) now becomes

(64a) $$\mathcal{N}\tilde{\theta}^1_0 = (\sigma + k^2)[\sqrt{2}\, e^{\sqrt{2}x};\ 0],$$

(64b) $$\mathcal{N}\tilde{Y}^1_0 = -(\sigma + k^2)[\sqrt{2}\, e^{\sqrt{2}x};\ 0].$$

These equations are to hold for $x \neq 0$, and zero boundary conditions apply at $\pm\infty$. Moreover, we obtain from (33) that zero boundary conditions hold at $x = 0$ as well. Let

(65) $$\mu_\pm \equiv 2^{-1/2}[1 \mp \Gamma], \qquad \Gamma \equiv \sqrt{1 + 2(\sigma + k^2)}.$$

Then the solution of (64) under the boundary conditions stated is

(66a) $$\tilde{\theta}^1_0 = 2^{1/2}[e^{\sqrt{2}x} - e^{\mu_- x};\ 0],$$

(66b) $$\tilde{Y}^1_0 = -2^{1/2}[e^{\sqrt{2}x} - e^{\mu_- x};\ 0].$$

I_0^1:
Adding (31a) and (31b), we have

$$\partial_{\xi\xi}(t_0^1 + y_0^1) = 0,$$

and from this and (46),

$$y_0^1 = t_{0+}^1 + \partial t_{0+}^1 \xi - t_0^1.$$

But (33), (41) imply

$$\partial t_{0+}^1 = 0, \tag{67}$$

so

$$y_0^1 = t_{0+}^1 - t_0^1. \tag{68}$$

Putting this into (31a) yields

$$\mathcal{M} t_0^1 = -t_{0+}^1 e^{t_0^0}.$$

Using (56) again, we obtain an equation analogous to (57)

$$\partial_\xi (p \partial_\xi t_0^1 - t_0^1 \partial_\xi p) = -t_{0+}^1 \partial_\xi e^{t_0^0},$$

which may be integrated to obtain

$$p \partial_\xi t_0^1 - t_0^1 \partial_\xi p = -t_{0+}^1 (e^{t_0^0} - 1).$$

We now let $\xi \to -\infty$ to obtain

$$\partial t_{0-}^1 = 2^{-1/2} t_{0+}^1. \tag{69}$$

From (41), the identity $2^{1/2} - \mu_- = \mu_+$, and (66a), we now obtain

$$\partial t_{0-}^1 = \partial \theta_0^1(0-) = \phi 2^{1/2} \mu_+. \tag{70}$$

Hence from (68),

$$\partial y_{0-}^1 = \phi \partial \tilde{Y}_0^1(0-) = -2^{1/2} \mu_+ \phi, \tag{71}$$

and from (69), (70),

$$t_{0+}^1 = 2\mu_+ \phi. \tag{72}$$

Moreover, (68), (42), (36) imply

$$0 = y_{0-}^1 = t_{0+}^1 - t_{0-}^1, \qquad t_{0-}^1 = 2\mu_+ \phi. \tag{73}$$

In all, we have

$$t_0^1 \simeq \phi 2^{1/2} \mu_+ \xi + 2\mu_+ \phi \quad (\xi \to -\infty), \tag{74a}$$

$$t_0^1 \simeq 2\mu_+ \phi \quad (\xi \to \infty), \tag{74b}$$

$$y_0^1 \simeq -2^{1/2} \mu_+ \xi \phi \quad (\xi \to -\infty). \tag{74c}$$

From (41) and (42) again, we may now identify

$$\theta_1^1(0-) = \theta_1^1(0+) = 2\mu_+ \phi. \tag{75}$$

O_1^1:
We will first consider this problem for the range $x<0$. Introduce $\tilde{\theta}_1^1$, $\tilde{Y}_1^1$ as in (63) and substitute into (26). From (53), (60), and (66b), we get

(76a) $$\mathcal{N}\tilde{\theta}_1^1 = -Q(\sigma+k^2)(2^{1/2}x+1)e^{\sqrt{2}x},$$

$$\mathcal{N}\tilde{Y}_1^1 = (\sigma+k^2)(Q-2^{1/2}l)(2^{1/2}x+1)e^{\sqrt{2}x} + l[-(\partial_{xx}-k^2)\tilde{Y}_0^1 - k^2\partial_x Y_0^0]$$

(76b) $$= [(\sigma+k^2)(Q-2^{1/2}l)+2^{3/2}l+(\sigma+k^2)(Q-2^{1/2}l)2^{1/2}x]e^{\sqrt{2}x} + l2^{1/2}(k^2-\mu_-^2)e^{\mu_- x}.$$

For our purposes, it will suffice to obtain the combination

$$Z \equiv \tilde{\theta}_1^1 + \tilde{Y}_1^1.$$

Adding (76a) and (76b), we obtain (still for $x<0$)

(77) $$\mathcal{N}Z = 2^{1/2}l[-(\sigma+k^2)(1+2^{1/2}x)e^{\sqrt{2}x} + 2e^{\sqrt{2}x} + (k^2-\mu_-^2)e^{\mu_- x}].$$

The solution of (77) under the appropriate boundary conditions is

(78) $$Z = l[-2^{1/2}-2x]e^{\sqrt{2}x} + [a_{1-}^1 + 2^{1/2}l - l(k^2-\mu_-^2)\Gamma^{-1}x]e^{\mu_- x} \quad (x<0),$$

where by (75)

(79) $$a_{1-}^1 \equiv \tilde{\theta}_1^1(0-) = 2\mu_+ = a_{1+}^1.$$

For the other range $x>0$, it is easy to see that

(80) $$Z = a_{1+}^1 e^{\mu_+ x} = 2\mu_+ e^{\mu_+ x} \quad (x>0).$$

I_1^1:
Adding (32a) and (32b), with use made of the relations (68), (48), we obtain

$$\partial_{\xi\xi}(t_1^1 + y_1^1 + lt_0^1) = 0;$$

hence (brackets denoting the change in the variable concerned from $\xi=-\infty$ to $\xi=\infty$)

(81) $$[\partial_\xi(t_1^1+y_1^1)] = -l[\partial_\xi t_0^1] = \phi 2^{1/2}\mu_+ l = \phi(1-\Gamma)l,$$

by (67) and (69). But we also have from (43), (44), and (66) that

(82) $$[\partial_\xi(t_1^1+y_1^1)] = \phi[\partial_x Z(0+) - \partial_x Z(0-)].$$

The right side is evaluated from (78), (79), (80), and the identities

$$\mu_1^2 = 1+\Gamma+\sigma+k^2, \qquad 2\mu_-\mu_+ = 1-\Gamma^2$$

to be

$$\phi\left[2\mu_+^2 - 1 + \Gamma^2 - l\left(-2+\frac{1+\sigma}{\Gamma}+\Gamma\right)\right].$$

Equating this to the right side of (81) gives the dispersion relation

(83) $$2\Gamma^2(\Gamma-1) = l(\sigma-\Gamma+1).$$

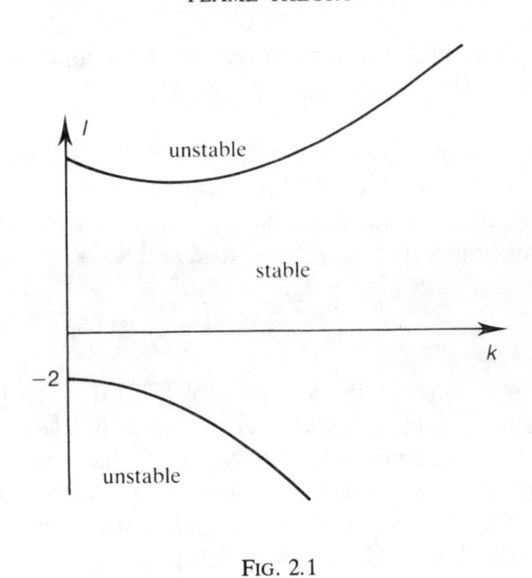

FIG. 2.1

This is a necessary condition relating l, k, and σ. Here k is required to be real, but σ could be complex.

2F. Stability. As mentioned, (83) is a necessary relation among k, σ, and l. If, for any given k and l, there is a solution for which $\operatorname{Re}\sigma > 0$, then the flame front is considered to be (linearly) unstable with respect to that wave number k.

By this criterion, the stability region is calculated to be as shown in Fig. 2.1. Passage from the stable region down through the lower boundary involves σ (real) passing through zero. For a fixed k, this would produce, by the usual bifurcation analysis, a new steady solution of (19), (20) with periodic dependence on z with wave number k. The plane flame front is thereby replaced by a "cellular" one. On the other hand, passage up through the upper curve corresponds to $\operatorname{Re}\sigma$ increasing through zero, $\operatorname{Im}\sigma$ at the same time remaining nonzero. The new solution this time by Hopf bifurcation will be periodic in both time and space. In fact, the latter kind of bifurcation also holds with $k = 0$, in which case it produces a planar time-periodic flame.

Presumably, if (83) holds, then the perturbative analysis could be continued to arbitrarily high order in ϵ and δ, but this has not been verified.

Along with the linear stability analysis described above, another natural and very important question is, "What alternate dynamics govern the motion of the flame front near places where stability is breached?" The most interesting such region is for l slightly less than its lower stability boundary of -2, and small k. Let $l = -2 - \eta$, $0 < \eta \ll 1$. Then the terms in (83) scale like $k = \kappa\sqrt{\eta}$, $\sigma = s\eta^2$, κ and s being $O(1)$ quantities, and to dominant order, the dispersion relation becomes

$$s = 8\kappa^2 - \kappa^4. \tag{84}$$

In this region, it is therefore natural to rescale space and time according to

$$z = \eta^{-1/2}\hat{z}, \quad \tau = \eta^{-2}\hat{\tau}.$$

If we do this and perform a weakly nonlinear stability analysis for $\phi = \psi(\hat{z}, \hat{\tau})$, we obtain, under still another $O(1)$ rescaling, the Kuramoto–Sivashinsky equation. We have reverted to the same symbols z and t to denote certain new rescaled space and time variables similar to $\hat{z}$ and $\hat{\tau}$:

$$\frac{\partial \psi}{\partial t} + \frac{\partial^4 \psi}{\partial z^4} + \frac{\partial^2 \psi}{\partial z^2} + \frac{1}{2}\left(\frac{\partial \psi}{\partial z}\right)^2 = 0.$$

This equation was derived by Sivashinsky [Si77a], [Si80], and in another context by Kuramoto and Tsuzuki [KT75], [KT76], [Ku78]. (For another review of this and related matters, see [MM].) It has remarkable properties, the most important being the existence of a finite-dimensional inertial manifold. It has been investigated extensively, both numerically and analytically (see, e.g., [HNZ] and references therein).

3. Flames due to chain branching.

There are two diffusive mechanisms that cause a flame to propagate. One is the diffusion, ahead of the flame, of some of the heat generated by its reactions. This heats the unburned gases in a neighborhood of the flame, bringing them up to the ignition temperature range. They then combust, causing the chain reaction that propagates the flame. The other mechanism is the diffusion ahead of the flame of certain radicals, again generated by the combustion process. These trigger chain-branching reactions which are essential to the reactive process. Radicals are short-lived compounds produced in most flames and essential to the combustion process, but generally not present for long in the combustion product. An example is the radical H (hydrogen atom, as opposed to the hydrogen molecule H_2), produced in hydrocarbon combustion, which triggers the reaction

$$H + O_2 \rightarrow O + OH.$$

This reaction produces two radicals O and OH for every one (H) it consumes; hence it is of the chain-branching variety. The extra radicals eventually recombine by reactions such as $H + H + M \rightarrow H_2 + M$, for example ($M$ being any third body that happens to be present, such as a molecule of H_2O).

Section 1 gives the main analytical details for simple flames propagating purely by thermal diffusion. Here we briefly discuss the analogous problem for an idealized flame propagating purely by chain-branching effects. The simplest model involves the single reaction

$$A_1 + A_2 \rightarrow 2A_2,$$

A_2 being a generic radical and A_1 the fuel being consumed. For further simplicity, assume that the rate does not depend on temperature, so that the temperature effects may be decoupled. Then in the system analogous to (5), we may ignore the first equation, and the second becomes a pair of equations for the concentrations Y_1 of A_1 and Y_2 of A_2. The scaled reaction rate there will be $\omega(Y) = Y_1 Y_2$, so we obtain

$$D_1 \partial_{xx} Y_1 + v \partial_x Y_1 - Y_1 Y_2 = 0, \tag{85}$$

$$D_2 \partial_{xx} Y_2 + v \partial_x Y_2 + Y_1 Y_2 = 0. \tag{86}$$

The unburned state has no radical, and if we call the initial fuel concentration unity, this system should be solved under the boundary conditions

$$Y_1(-\infty) = 1, \quad Y_2(-\infty) = 0, \quad Y_1(\infty) = 0, \quad Y_2(\infty) = 1. \tag{87}$$

In the special case $D_1 = D_2 = D$, adding the two equations produces a single equation for $W \equiv Y_1 + Y_2$:

$$D \partial_{xx} W + v \partial_x W = 0,$$

the only bounded solutions of which are constants. The particular constant in this case is the one obtained from (87) by evaluating W at $-\infty$, namely 1: $W \equiv 1$. Hence $Y_1 = 1 - Y_2$, and (86), (87) become

$$D \partial_{xx} Y_2 + v \partial_x Y_2 + Y_2(1 - Y_2) = 0, \tag{88}$$

$Y_2(-\infty) = 0$. At $+\infty$ the solution should approach the only other critical point, which is $Y_2(\infty) = 1$.

Equation (88) is the traveling wave equation for the well-known Fisher's equation (see [Fis], [AW75], [AW78], or [Fi79c], for example). It does have solutions; in fact it has an infinite number of them, one for every possible velocity v satisfying

$$v \geq v_0$$

for a certain positive "minimal velocity" v_0. As shown in [AW78] in an entirely different context, the solution of the Fisher traveling wave equation representing propagation of a wave into a region that was initially totally free of the radical ($Y_2 = 0$ initially) is the one with minimal velocity v_0. This wave, then, would be the relevant one in the flame context.

Incidentally, there is no cold boundary difficulty in this problem because the unburned state is in (unstable) equilibrium. It is drawn out of that equilibrium by the radicals diffusing ahead of the main part of the flame.

4. Complex chemistry.

Flames typically involve a large number of reactions and species. This complexity makes it very difficult to comprehend on an intuitive level the essential character of the combustion events. Moreover, computations are expensive and sometimes the reasons for the phenomena they display are not immediately enlightening. It is therefore extremely important, if possible, to systematically reduce the complex problem to understandable terms. For example, we would like to know (i) which of the reactions are really important, and why; (ii) what the typical orders of magnitude of the various species, especially the radicals, are during the combustion process; and (iii) whether the essential features of the flame can be deduced from a simpler skeleton model with many fewer reactions and species.

There have been several approaches used in the past to move toward this goal. Sensitivity analysis, for example, is a very popular avenue to singling out the important reaction paths (for an example of this, see [SRRD]). Other kinds of information have been used to reduce complex chemical kinetics to simpler proportions, as in [Cl], [PLP], [Pe], [PK], [PS], [PW]. Finally, under the assumption that at least one reaction has high activation energy and that for all temperatures the various reaction constants $k(T)$, except for two or three of them, are sufficiently far apart from one another, the scheme proposed by Fife and Nicolaenko [FN83], [FN84a], [FN84b] is then applicable and provides a way to gauge the importance of the various reactions, along with the spatial structure of the flame and the order of magnitude of its velocity.

The last of the three types of approaches mentioned is the most systematic, but it cannot be applied in typical cases when there are many reactions with rate constants that are close to each other. It was the desire to understand such flames, for example those formed from the oxidation of hydrogen, that motivated the recent development of the approach that will be described here. It is designed to give rough qualitative answers to the above questions in some cases when the assumptions of the Fife–Nicolaenko scheme are not fulfilled. The exposition follows [Fi88].

This approach is based first on the assumption that there is a thin flame layer at a characteristic temperature T in which either all or part of the reaction network under consideration comes to chemical equilibrium. Such a layer would be occasioned by some reaction(s) with high activation energy which triggers the other reactions taking place in the layer. If only part of the network does this, then the rest of the network may come to equilibrium in another (not necessarily localized) site on the flame profile at a slower rate.

Second, it is assumed that each reactant has a characteristic average level of concentration in the layer.

Third, it is based on often exaggerating the differences in magnitudes of rates and concentrations. Exaggeration is the essence of such asymptotic models. Consequently the results obtained are expected to be only qualitatively accurate in many cases. Nevertheless, they provide a clear picture of the main chemical processes at work to produce the flame.

One form of this model was applied in [Fi, pr] to $H_2 - O_2$ flames using a 19-reaction network. The analysis there was done by hand and encouraging results were obtained. Later in [RF], [Ro] the procedure was implemented on a computer using the language of MACSYMA. At the same time, an interpretation of the procedure in terms of vertices of polytopes was given and the concept of critical temperatures developed. The latter are temperatures at which the geometry of crucial vertices changes. This scheme has been used to categorize flames resulting from the combustion of ozone by Fife and Gill. For other approaches to the ozone system, see [RLW] and [RW].

Here we give a more general formulation of the model and explain it in the context of suggestive simple examples.

4A. The framework. For convenience we operate within the framework of a thermodiffusive model, although a more general context will lead to the same traveling wave equation (90) below. First, we will develop the equations for steady plane flames governed by a general network $\mathcal{M}$ of m possibly reversible reactions,

$$\mathcal{M} = \{R1^+, R1^-, R2^+, \cdots, Rm^-\}. \tag{89}$$

Those equations take the form (as in [FN83])

$$DU'' - MU' + K \cdot (\boldsymbol{\omega}^+(U) - \boldsymbol{\omega}^-(U)) = 0. \tag{90}$$

The meaning of the symbols is as follows. The independent variable is the traveling wave coordinate $x = \bar{x} + v\tau$. The $(n+1)$ vector U is defined as

$$U = \begin{bmatrix} T \\ Y_1 \\ \cdots \\ Y_n \end{bmatrix} \tag{91}$$

where T is temperature and Y_i are the concentrations. We nondimensionalize T so that one unit of T is the difference ΔT between a typical burned temperature T_+ in the flame and the unburned temperature T_-. The concentration Y_i will be defined as the mass fraction of species A_i in the gas at any position and time divided by the molecular weight of that species.

The constant M is the unknown rescaled mass flux. D is a diffusion matrix, assumed to be constant. The wave moves to the left, so that the unburned conditions far upstream are attained at $x = -\infty$: $U(-\infty) = U_-$ (say).

The $(n+1) \times (m)$ stoichiometric matrix K is constructed directly from the sets of reactions Rj and their heat releases. First, a matrix K^+ is defined by setting each element κ_{ij}^+ equal to the stoichiometric coefficient of A_i on the left side of reaction Rj. These coefficients are those appearing in the representation of Rj as

$$\Sigma \kappa_{ij}^+ A_i \rightleftarrows \Sigma \kappa_{ij}^- A_i.$$

Thus, for example, if $R4$ is $A_2 + A_3 \to 2A_1$, then $\kappa_{24}^+ = \kappa_{34}^+ = 1$ and all the other elements in column 4 with $i \geq 1$ are zero. There is also a zeroth row in the matrix, with $\kappa_{04}^+ = -Q_4$, a suitable measure of the heat released by reaction $R4$. This completely defines the matrix K^+. The matrix K^- is defined in the same way by looking at the right sides of the reactions. However, the zeroth row of K^- is set equal to zero. Finally,

$$K \equiv K^- - K^+. \tag{92}$$

The m-vectors $\boldsymbol{\omega}^+$ and $\boldsymbol{\omega}^-$ give the rates of the forward and reverse reactions, respectively. The rates are assumed to be of mass action type

$$\omega_j^{\pm}(U) \equiv k_j^{\pm}(T) \prod_i Y_i^{\kappa_{ij}^{\pm}}. \tag{93}$$

The products are mass action monomials, and the k's are reaction constants, generally determined (if indeed possible) by experimental measurements.

This defines all the symbols in the basic equation (90).

4B. The model problem. It is assumed that part (at least) of the original network $\mathcal{M}$ (call it $\mathcal{M}'$) comes to chemical equilibrium in a thin layer (primary reactive zone) at a characteristic temperature T. The rest of the network may also come to equilibrium, but this is assumed to happen at a slower rate and elsewhere in the flame profile. This other part of the profile will have a larger characteristic length scale. The profile therefore will be relatively flat downstream of the primary combustion layer.

In view of that, we may integrate (90) with respect to x from $-\infty$ to the relatively flat region immediately downstream of the layer. The integral of U'' vanishes approximately because of the flatness assumption. Recall that U_- is the unburned state far to the left (where the integration begins). Let $U_+ = (T_+, Y_+)$ be the state attained by the chemical processes in the layer, so that $U = U_+$ at the end of the integration region behind the layer. Then we obtain

$$Y_+ = Y_- + K \cdot \boldsymbol{\alpha}, \tag{94}$$

where the "allocation" vector $\boldsymbol{\alpha}$ is defined by

$$\boldsymbol{\alpha} \equiv \frac{1}{M} \int (\boldsymbol{\omega}^+(U(z)) - \boldsymbol{\omega}^-(U(z)))\, dz. \tag{95}$$

In the same way, the integrated zeroth component of (90) yields

$$T_+ = T_- + Q \cdot \boldsymbol{\alpha}, \tag{96}$$

where Q is the vector of heat releases.

In accordance with the second point listed in the first paragraph of this section (§4), we associate with each concentration Y_i a characteristic average order of magnitude in the reaction layer

$$Y_i \simeq e^{-\nu_i} \leq 1, \tag{97}$$

where $\nu_i \geq 0$. (Actually, we also want to ensure that at least one of the Y's is an $O(1)$ quantity, so we may normalize them to achieve that purpose, as was done in [Fi, pr], [RF].)

We are interested in examining the relative magnitudes of the ω's and do this by taking the negative natural logs of the expressions (93), and subtracting an (unknown) parameter s which is intended to be chosen so that the minimum of the resulting expressions is equal to zero. (This is not quite true when "partial equilibrium" conditions exist, as we will see later.) We thus define

$$\beta_j^{\pm}(\mathbf{v}, s, T) \equiv l_j^{\pm} + \Sigma \kappa_{ij}^{\pm} v_i - s. \tag{98}$$

In this expression, the l's are negative logarithms of the k's, and are assumed to be given nonincreasing functions of T. The summation is over i. Of course $\mathbf{v} \equiv (v_1, v_2, \cdots, v_n)$.

For a given Y_-, we attempt to judge the relative importance of the reactions at a given temperature T (in the flame layer) and at the same time the magnitudes of the concentrations (measured by $\mathbf{v}$) in the following way. We choose a nonnegative resolution parameter ρ, and seek values of $\mathbf{v}$, s, and $\boldsymbol{\alpha}$ such that

(99a) (i) for each j, $\beta_j^{\pm} \geq 0$ or $\beta_j^+ = \beta_j^- < 0$,

(99b) (ii) $\mathbf{v} \geq 0$,

(99c) (iii) $\beta_j^+ > \rho \Rightarrow \alpha_j \leq 0$; $\beta_j^- > \rho \Rightarrow \alpha_j \geq 0$,

(99d) (iv) with Y_+ defined in terms of $\boldsymbol{\alpha}$ by (94), $Y_+ \geq 0$,

(99e) (v) $v_j > \rho \Rightarrow Y_{+j} = 0$,

(99f) (vi) Y_+ is in equilibrium for the subnetwork $\mathcal{M}' = \{Rj: \beta_j \leq \rho\}$.

The first part of (i) is the role of the normalization parameter s. The second part will be explained later when partial equilibrium is discussed. The requirement (ii) was mentioned before. Requirement (iii) expresses the fact that large β's mean small rates; hence by (95) the corresponding α's (which, by the way, are all $\leq O(1)$ because of (94)) are approximated by zero. If the rate of a reaction is small in only one direction, then again by (95) the corresponding $\boldsymbol{\alpha}$ will be of predetermined sign. Requirement (iv) is obvious; (v) simply expresses the approximation that when v_j is large, Y_j is small, and we approximate it in the final state by 0. Finally, (vi) says that by excluding those reactions whose rates are so small they can be neglected anyway, the network goes to equilibrium at the end of the primary burning process. If it did not, the burning would not be complete. A later combustion process in a larger region behind the primary layer may take care of those remaining reactions that were first neglected. Reaching equilibrium here means reaching a state U_+ such that the reaction term in (90) vanishes:

$$\tilde{K} \cdot \boldsymbol{\omega}^+(U_+) - \boldsymbol{\omega}^-(U_+)) = 0, \tag{100}$$

the matrix $\tilde{K}$ denoting the stoichiometric matrix of the reduced network $\mathcal{M}'$.

In seeking solutions of (99), the strategy will be to concentrate on solutions for which as many as possible of the inequalities in (99a), (99b) are equalities.

The reason is most easily seen in the case where we choose $\rho = 0$ (this bold choice turns out to give meaningful results and was often used in [Ro], [RF]). Then because of (99c), the inequalities β_j^+, $\beta_j^- > 0$ force $\alpha_j = 0$, which effectively eliminates Rj from $\mathcal{M}'$. Likewise by (99c), $\nu_j > 0$ imposes an additional constraint on the α's. Our strategy, therefore, leaves as few as possible constraints on the α's; hence it gives a wider range of possibilities for the participating network. On the one hand, this is the most secure strategy in that it better provides that viable possibilities are not inadvertently missed. On the other hand, it may well result (and in fact often does) in the existence of a nontrivial solution of (99), which might not exist away from the vertices; freedom in the choice of α's will better provide for the existence of allocations satisfying (99d)–(99f).

Notice that the temperature T_+ of the layer where we are doing our analysis depends on $\boldsymbol{\alpha}$. We will see later that this coupling generates some interesting phenomena.

There is no guarantee that, for a given T, a solution of (99) exists, and likewise there is no guarantee that it is unique if it does exist. The consequences and possibilities of nonuniqueness are under current investigation. We feel that nonexistence generally implies a kinetic extinction phenomenon.

Finally, the obvious crudeness of the model can be alleviated by passing to a second approximation, as will be explained below.

We illustrate the process by two simple examples.

4C. Example 1. The sequential reaction $A_1 \to A_2 \to A_3 \to 0$. We denote the three reactions, proceeding from left to right, by $R1$, $R2$, and $R3$. We suppose that k_2 and k_3 do not depend on temperature, but that k_1 does very strongly (has high activation energy). The initial state is

$$Y_- = \begin{bmatrix} 1 \\ 0 \\ 0 \end{bmatrix}.$$

The following rough picture is fairly clear. The high activation energy of the first reaction causes a primary layer like the classical one for a one-step reaction, at a specific temperature T.

(a) If, at the given temperature T, $k_1(T)$ is much smaller than k_2 and k_3, then $R2$ rapidly consumes any amount of A_2 produced by $R1$, and its product A_3 is again rapidly consumed by $R3$.

(b) If, on the other hand, $k_1(T)$ is much larger than k_2, then all A_1 is consumed in a primary flame layer in which $R1$ goes to completion (i.e., the subnetwork consisting only of $R1$ comes to equilibrium). The remaining reactions $R2$ and $R3$ are not yet in equilibrium, but approach it at a much slower rate behind the primary layer (notice this is true even if k_3 is very large; no assumption is being made about its magnitude). Those reactions are spread out over a much larger domain.

(c) If k_1 is much smaller than k_2 but much larger than k_3, then $R1$ and $R2$ quickly come to completion in the primary layer, and $R3$ does so later in a trailing zone.

These results are trivial to grasp and we do not need any complex machinery to deduce them. However, the systematic procedure described in the previous section is necessary for more involved networks in which the results are not so intuitive, and so it will be worthwhile to illustrate that procedure in the context of this simple example. Moreover, this analysis will clarify some of the points in the above categorization, including some of the assumptions made. We therefore proceed as outlined in the previous section.

Clearly $K^- = 0$ and

$$K = \begin{bmatrix} Q_1 & Q_2 & Q_3 \\ -1 & 0 & 0 \\ 1 & -1 & 0 \\ 0 & 1 & -1 \end{bmatrix}. \tag{101}$$

We will need to estimate the width of the flame layer. According to the standard theory of a one-step high activation energy reaction [BL82], the ratio of the width of the preheat zone to that of the flame layer is roughly equal to the Zel'dovich number

$$\theta \equiv \frac{d}{dT} \ln k_1(T),$$

this expression being evaluated at the layer's temperature T. Moreover, relative to the width of the preheat zone, the slope of the concentration profile $Y_1(x)$ will be an $O(1)$ quantity. Given that $Y_1(x)$ also vanishes at the back of the layer, it follows that its average value in the layer itself will be $O(\theta^{-1})$. Therefore, a priori from (97) we may set

$$\nu_1 = \ln \theta.$$

The other concentration parameters ν_2 and ν_3, as well as the parameter s, are not known beforehand, however.

Recalling $l_j = -\ln k_j$, we have by (98)

$$\beta_1 = l_1 + \nu_1 - s, \quad \beta_2 = l_2 + \nu_2 - s, \quad \beta_3 = l_3 + \nu_3 - s. \tag{102}$$

From (94) we also have

$$Y_+ = \begin{bmatrix} 1 - \alpha_1 \\ \alpha_1 - \alpha_2 \\ \alpha_2 - \alpha_3 \end{bmatrix}. \tag{103}$$

Since there are no reverse reactions, the second part of (99a) can be overlooked, and (99a) says

$$s \leq l_1 + \nu_1, \quad \nu_2 \geq s - l_2, \quad \nu_3 \geq s - l_3. \tag{104a}$$

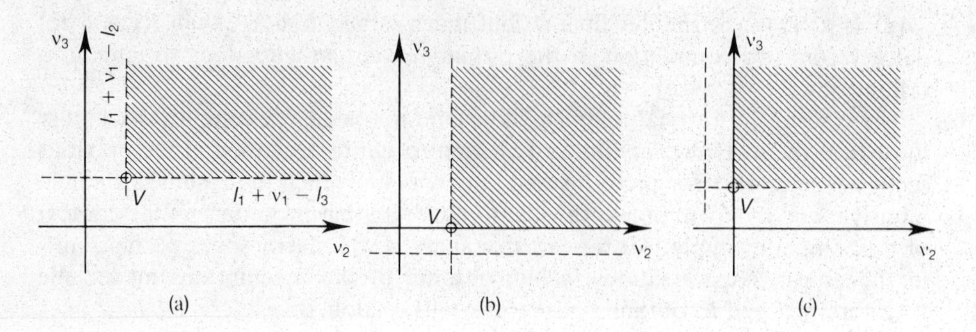

FIG. 2.2. *Allowed parameter regions for the first example, cases* (a), (b), *and* (c).

Condition (99d), with (103), says that

(104b) $$\alpha_1 \leq 1, \quad \alpha_2 \leq \alpha_1, \quad \alpha_3 \leq \alpha_2.$$

Therefore, in order for any combustion to occur at all, we have to have $\alpha_1 > 0$. So from (99f), reaction $R1$ must go to completion, i.e., $Y_{1+} = 0$; hence from (103), $\alpha_1 = 1$. Let us take the resolution constant ρ to be zero, a very conservative stance. Then since $\alpha_1 > 0$, we must have $\beta_1 = 0$ from (99c). Hence from (102) and (104a),

(105) $$s = l_1 + v_1, \quad v_2 \geq l_1 + v_1 - l_2, \quad v_3 \geq l_1 + v_1 - l_3.$$

We consider four cases.

(a) $l_1 + v_1 > l_2$, $l_1 + v_1 > l_3$. Then the region allowed by (105) in the positive quadrant of the (v_2, v_3) plane is as shown in the shaded part of Fig. 2.2(a). It has one vertex V, defined by making the inequalities in (105) equalities. We test the vertex of the region to see if it satisfies the rest of (99). (We do this in general, working only at the vertices, in accordance with the strategy given at the end of §4B.) The vertex V has all β's equal to zero, so all reactions participate, and all α's are positive. Hence by (99f), the entire network must come to equilibrium, which means that $Y_+ = 0$. From (103) and equalities in (105), we get

$$\boldsymbol{\alpha} = (1, 1, 1), \quad \mathbf{v} = (0, l_1 + v_1 - l_2, l_1 + v_1 - l_3), \quad s = l_1 + v_1.$$

This is the only solution of (99) in this case.

(b) $l_1 + v_1 > l_2$, $l_1 + v_1 < l_3$. The region defined by (105) and (99b) in this case is as shown in Fig. 2.2(b). The only vertex is at $(l_1 + v_1 - l_2, 0)$. Now $\beta_3 > 0$, so $\alpha_3 = 0$ by (99c). But $\beta_2 = 0$, so by (99f) in fact $R1$ and $R2$ must pass to equilibrium, meaning $Y_{1+} = Y_{2+} = 0$ or $\alpha_1 = \alpha_2 = 1$. Thus

$$\boldsymbol{\alpha} = (1, 1, 0), \quad \mathbf{v} = (0, l_1 + v_1 - l_2, 0), \quad s = l_1 + v_1.$$

This is the only case satisfying (99). There is a residual amount of A_3 left over after the primary layer.

(c) $l_1 + \nu_1 < l_2$, $l_1 + \nu_1 > l_3$. Then the vertex (Fig. 2.2(c)) is at $(0, l_1 + \nu_1 - l_3)$, which means that $\beta_3 = 0$ but $\beta_2 > 0$. The latter says that $\alpha_2 = 0$; hence by (104b), $\alpha_3 = 0$ as well. Thus $\boldsymbol{\alpha} = (1, 0, 0)$, $\boldsymbol{\nu} = (\nu_1, 0, l_1 + \nu_1 - l_3)$, $s = l_1 + \nu_1$. This is the only solution of (99). We could then argue that in fact $\alpha_2 = 0$ implies there is no production of A_3, so we should have $\nu_3 = \infty = \beta_3$. This shifts the above-named vertex to one at ∞, namely $(0, \infty)$. But in view of our conservative choice $\rho = 0$, this does not change the original conclusion that A_3 exists in a negligible amount, since $\nu_3 > 0$.

(d) $l_1 + \nu_1 < l_2$, $l_1 + \nu_1 < l_3$. The analysis in this case is essentially the same as that in (c).

These conclusions reproduce the original easy intuitive conclusions if inequalities such as "$l_1 + \nu_1 < l_2$" are interpreted to mean that "k_1 is *much* greater than k_2." Of course it does not really imply the latter, but this only emphasizes the fact that our qualitative model exaggerates differences.

4D. Example 2. The chain-branching, chain-terminating network $\mathbf{A_1 + A_2 \rightleftarrows 2A_2}$, $\mathbf{A_2 \to 0}$. Take the initial concentration to be $Y_- = \left[\begin{smallmatrix}1\\0\end{smallmatrix}\right]$, and the first forward reaction to have high activation energy, so that when the initial supply of A_1 is consumed, it will be in a thin layer with characteristic temperature T. For simplicity, let us assume the other reactions have zero activation energy.

As in the previous example, we first give an intuitive account. If the reaction rate $k_2(T)$ of the second reaction is much larger than $k_1^+(T)$, then $R2$ outcompetes $R1^+$ for any available supply of A_2. Since there is none to begin with, no combustion can be sustained and there is no flame.

Now suppose k_2 is much smaller than k_1^+. Then $R1$ comes to equilibrium on a time scale that is relatively short compared to that associated with $R2$. This first equilibration may be the only thing that occurs in the primary layer. The equilibrium values of Y_1 and Y_2 are related by

$$k_1^+ Y_1 = k_1^- Y_2.$$

If k_1^-/k_1^+ is large, then Y_2 is small, meaning that not much A_2 has been produced; we therefore have α_1 small and we approximate it by 0. Again, the only possibility for any primary flame would be $\boldsymbol{\alpha} = 0$, and we conclude that in fact there is no primary layer. On the other hand, there will be a small amount of Y_2 (of order k_1^+/k_1^-) present, and so $R2$ will gradually operate to use up all the A_2 produced. It is essentially produced from A_1 through $R1^+$, and so the initial supply of A_1 will be slowly consumed. This will not occur in a layer. In fact, the heat release from $R2$ in general will cause the temperature to rise gradually during the process, and there will be no one characteristic temperature.

On the other hand, if k_1^-/k_1^+ is small then the first equilibration event occurs in a layer and converts most of the initial supply of A_1 into A_2. This is the primary layer; the new supply of A_2 is then slowly consumed by $R2$.

Now let us corroborate these quite easy conclusions by use of the formalism

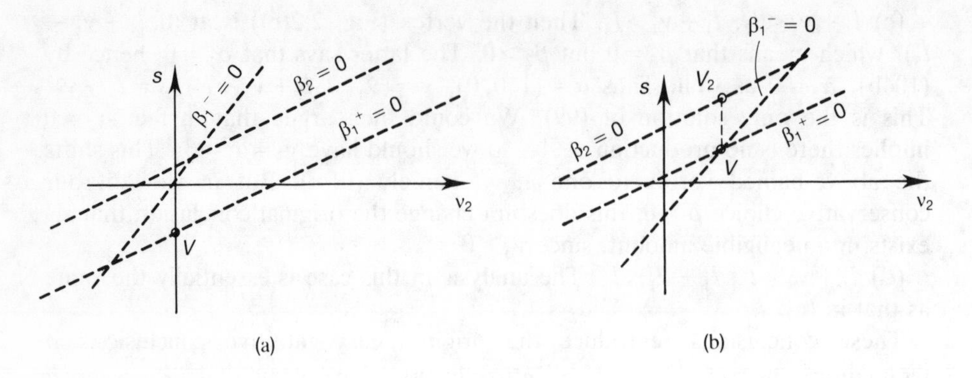

FIG. 2.3. *Allowed parameter regions and vertices for the second example.*

of (99). As before, we argue that if combustion really occurs, then $R1^+$ must indeed go to completion in a layer, and therefore that we may set $\nu_1 = \ln \theta$ on an a priori basis. Then from (98) and (94)

$$\beta_1^+ = l_1^+ + \nu_1 + \nu_2 - s, \quad \beta_1^- = l_1^- + 2\nu_2 - s, \quad \beta_2 = l_2 + \nu_2 - s, \tag{106}$$

$$Y_{1+} = 1 - \alpha_1, \qquad Y_{2+} = \alpha_1 - \alpha_2. \tag{107}$$

Case I. $l_2 < l_1^+ + \nu_1$. Then from (106) and (99a) we have that $\beta_1^+ > 0$ always, so by (99b), $\alpha_1 \le 0$. But (107) and (99c) imply $\alpha_2 \ge 0$ and $\alpha_1 - \alpha_2 \ge 0$, so in all, $\boldsymbol{\alpha} = 0$. No combustion is possible.

Case II. $l_2 > l_1^+ + \nu_1$. In the (ν_2, s) plane the region defined by the inequalities stating that all β's and all ν's are ≥ 0 appears as either Fig. 2.3(a) (Case II(a)) or Fig. 2.3(b) (Case II(b)).

In Fig. 2.3(a), we have a vertex V at $\nu_2 = 0$, $s = l_1^+ + \nu_1$. At V, $\beta_2 > 0$ so by (99c) we set $\alpha_2 = 0$. We also have $\beta_1^- > 0$ so we neglect $R1^-$. Thus $R1^+$ must come to equilibrium ($\mathcal{M}' = \{R1^+\}$); hence $Y_{1+} = 0$ and by (107) $\alpha_1 = 1$. In short, the primary layer has

$$\boldsymbol{\alpha} = (1, 0), \quad \nu_2 = 0, \quad s = l_1^+ + \nu_1.$$

This, of course, will be followed by the slow consumption of A_2 outside the layer.

In Fig. 2.3(b), the analogous vertex V is at $\nu_2 = l_1^+ + \nu_1 - l_1^-$, $s = 2l_1^+ + 2\nu_1 - l_1^-$. Again, $\alpha_2 = 0$ for the same reason, but it can no longer be said that $\alpha_1 = 1$, because now $\beta_1^- = 0$. Also $\nu_2 > 0$, so by (99e) and (107) we must have $\alpha_1 = \alpha_2 = 0$, and there is no primary layer.

There is another vertex in Fig. 2.3(b), however, which we have so far overlooked. According to (99a), we must allow the possibility that $\beta_1^+ = \beta_1^- < 0$. This is shown as vertex V_2 in Fig. 2.3. At that vertex, $\beta_2 = 0$, so by (99f), $R2$ must come to equilibrium, meaning $Y_{2+} = 0$. By (107), then, $\alpha_1 = \alpha_2$. On the other hand, by (99f) the entire network must come to equilibrium. Equilibrium for $R1$ means that Y_{1+} and Y_{2+} are linearly related. Since the latter was just shown to be zero, the former must also be zero. Therefore by (107), $\alpha_1 = 1$. We conclude that the unique *nontrivial* solution of (99) is

$$\boldsymbol{\alpha} = (1, 1), \quad \nu_2 = (0, l_1^+ + \nu_1 - l_1^-), \quad s = l_2 + l_1^+ + \nu_1 - l_1^-.$$

Now we may see that the systematic treatment mirrors the intuitive treatment. In fact, Case I corresponds to the case when k_2 is much larger than k_1^+, Case II(a) corresponds to the case when it is much smaller than k_1^+, as is k_1^-, and Case II(b) corresponds to the remaining case when k_1^+ is much larger than k_2, but much smaller than k_1^-.

All of this leaves the question open of what happens in the intermediate cases, when the inequalities are not so extreme. Then of course the character of the chemistry is intermediate between those of the extreme cases. It is best and easiest first to delineate the extreme cases, and then proceed from there by modifying and moderating the results.

4E. The concept of partial equilibrium. In Fig. 2.3(b), we have $\alpha_1 = \alpha_2 = 1$, even though the rate of reaction $R1^+$, as measured by $\beta_1^+ < 0$, is much larger than that of $R2$, since $\beta_2 = 0$. This is no paradox, however; it simply means that the reverse reaction $R1^-$ also has large rate, and that rate partially balances the forward rate, so that the net rate of $R1$ is the same order of magnitude as that of $R2$. When this happens, the reaction $R1$ is said to be in partial equilibrium; it is a very common phenomenon in reaction kinetics. We could define the degree of partial equilibrium by the difference between $\beta_2 = 0$ and $\beta_1^+ = l_1^+ + \nu_1 - l_2$. When temperature varies by a small amount, causing l_1^+ to vary, this degree of partial equilibrium will also change so that $R1^+$ and $R1^-$ balance to a different extent, but the essential character of the flame will not change.

The rationale for the second part of stipulation (99a) is now clear. It is designed to account for reactions being in partial equilibrium.

4F. Parameter variation. It is interesting to study the changes in the solutions of (99) when one of the parameters in the problem (the most meaningful probably being T) changes. It is natural to characterize the vertex generating the solution by specifying which of the inequalities in (99a) and (99b) are actually equalities at the vertex. Even though the position of a vertex will move around in parameter space, let us agree that the vertex retains its identity if and only if that characterization does not change. Then as the one variable parameter changes, the corresponding vertex will only change at discrete values of the parameter, and then the change will be abrupt. Usually (as in these examples) $\boldsymbol{\alpha}$ will also remain constant, changing only when the vertex changes in the above sense. On the other hand, $\boldsymbol{\nu}$ of course will change continuously.

In Example 1, let us choose l_1 to be the variable parameter. We begin with large enough l_1 so that case (a) holds, and $\boldsymbol{\alpha} = (1, 1, 1)$. As l_1 is decreased, at a critical point case (a) changes into case (b) or case (c), depending on which is larger: l_2 or l_3. In the first case, $\boldsymbol{\alpha}$ suddenly shifts to $(1, 1, 0)$, the third reaction essentially being quenched in the primary layer, and in the second to $(1, 0, 0)$.

If all three reactions are exothermic and T_- is fixed, then the temperature T in the layer depends on $\boldsymbol{\alpha}$ according to (96), and so of course will change at the critical parameter point. Considering that k_1 and l_1 depend on T, we may find ourselves in a situation in which increasing T_- (say) will cause T to increase, hence causing l_1 to decrease until it reaches the critical point. At that point the allocation changes and because of that change, T suddenly decreases. This in turn raises l_1 again, which turns the old allocation back on, etc. Finally a compromise will result, in which the flame layer is host to a mixture of the two allocations in such a proportion that the temperature there is exactly the critical temperature. This critical temperature then will be insensitive to changes in T_- within a certain range. This is an example of a temperature plateau, as pointed out in [CFN] and [FN82].

It should be kept in mind, however, that this temperature plateau is only with regard to the primary layer. The final burned temperature of the gas will always be given by (96) with $\boldsymbol{\alpha} = (1, 1, 1)$, but it may be attained partly through combustion in the zone behind the layer.

A similar critical transition occurs in Example 2; in this case the change involves the onset of partial equilibrium in the primary layer.

Finally, the changes in the solution of (99) are abrupt, but they correspond to more gradual changes in any real flame being modeled. It has to be understood that the discontinuities in the model will be smoothed out in reality.

4G. Further approximations; Discussion. The procedure as described here is fraught with exaggerations and so may result in a crude approximation, but gives a clear qualitative picture. At the same time, the result may well serve as a basis for better approximations, as was shown in [Fi, pr]. Any viable solution of (99), in fact, corresponds to a vertex in $(\boldsymbol{\nu}, s)$ parameter space. At that vertex, some of the β's are $\leq \rho$ and result in nonzero allocation components. But all of the β's, whether eliminated on the first round or not, can be evaluated at that value of $(\boldsymbol{\nu}, s)$, and the extent of participation of the various reactions estimated. In general, it will be found that some of the reactions that were eliminated in the first approximation actually do participate to some (perhaps minor) extent. A further effect of this second approximation will be to "smooth out" the sharp changes that occurred at the critical temperatures mentioned above.

Solving the model problem as explained here gives a reduced reaction network consisting of only those reactions that have nonzero α's. This is expected to be a skeleton network giving the essential chemistry for the flame. However, the analysis is all temperature-dependent, and the skeleton network so obtained may very well change at critical temperatures. It therefore may be true that there is no single simplified network that will be valid for the entire temperature range of interest.

The reduced network obtained by this approach, of course, may imply overall effective chemistry through a further reduced network in which those radicals that do not persist behind the primary combustion zone are eliminated. However, this overall network will no longer satisfy mass action kinetics, and so will be of limited use. The rates of its reactions can be obtained from the β's in our calculations, but they may be rather complicated functions of temperature and concentrations.

As mentioned before, the allocation $\boldsymbol{\alpha}$ obtained in the primary layer itself determines the temperature of that layer through (96). On the other hand, the vertex of interest, hence $\boldsymbol{\alpha}$, was determined as a function of temperature in the layer. This poses an obvious compatibility problem: will the temperature computed by (96) be the same as the one with which we started? This question of compatibility produces interesting consequences. First of all, it should be realized that although the vertex does change continuously with temperature, $\boldsymbol{\alpha}$ does not. In fact, the α's satisfy certain linear constraints (99d), (99f) to which additional constraints (99e) have been added; the latter are the only constraints on the α's that depend on the ν's, hence on the temperature. They change only at critical temperature values. It therefore follows that $\boldsymbol{\alpha}$ will be constant between those critical temperature values. This makes the compatibility problem less serious than it may appear at first glance because the temperature computed from (96) now need be only in a certain interval.

On the other hand, examples in [RF] show that it may not be possible to get a compatible temperature this way, and the actual resulting flame temperature may be a compromise, such as was indicated in §4F. Then a temperature plateau results. Such plateaus were found in [RF]. Another possibility is nonuniqueness due to this temperature coupling effect, explored in [HaH], [CFN], [FN82]. Examples of this kind of nonuniqueness were not found in the H_2–O_2 system.

CHAPTER 3

Electrophoresis

Electrophoresis is a collection of various techniques used to separate ions (often charged organic molecules) in solution by means of their differing reactions to an imposed electric field.

Mathematical models in terms of nonlinear partial differential equations are given by Babskii et al. [BZY] and Saville and Palusinski [SaP], for example. The most relevant solutions of these equations will have layer structures, representing either abrupt transition zones where the concentrations of separated chemicals change or zones where they aggregate (as in isoelectric focusing). The formation and movement of these structures is of high interest.

1. Models and problems.

To develop the models here, we start by looking at an individual ion, carrying a charge of z electronic charge units. If it is in a unidirectional electric field of strength E, then it experiences a force and therefore by viscous effects a limiting velocity q proportional to Ez. Call the proportionality constant μ; it is called the mobility of that ion, and we assume it to be independent of everything except what species we are considering. Thus

$$q = \mu z E. \tag{1}$$

Now suppose that these ions exist in solution with a concentration u. Their collective motion will cause an induced flux f_0 given by

$$f_0 = \mu z E u. \tag{2}$$

To this we may add a diffusive flux obeying Fick's law: $f_d = -du_x$ with d, the diffusivity, again assumed constant. It is generally assumed that Einstein's relation holds, namely that d and μ are related by $d = \alpha\mu$, α being a constant independent of the species, depending, in fact, only on temperature. We now have a total flux

$$f = \mu z E u - \alpha \mu u_x. \tag{3}$$

Next, suppose there is a mixture of n different species in solution; there is a corresponding concentration vector $u = (u_1, \cdots, u_n)$, flux vector f, as well as charge vector z and mobility vector μ. E is still a scalar, since we are supposing space to be one-dimensional. Now the expression (3) still holds if we interpret products among the vectors u, μ, z to be *componentwise products*.

In addition, there may be chemical reactions among the various species, resulting in a source function $g(u)$, where $g: R^n \to R^n$. The set of conservation laws associated with this flux and source term is

$$u_t + f(u, u_x)_x = g(u),$$

i.e.,

$$u_t + \mu(z(Eu)_x - \alpha u_{xx}) = g(u). \tag{4}$$

In this system, E is also an unknown function (as well as u), since there will be a feedback influence, the concentration distribution u causing a charge distribution which in turn alters the field. This last effect is expressed by means of Poisson's equation

$$\beta E_x = -z \cdot u, \tag{5}$$

β being the ratio of permittivity of the solvent to the molar charge, and the right side representing the (negative of the) total charge density in the solution. On the right we have the scalar product between the two vectors z and u.

Our basic differential equations ($n+1$ of them in all) are (4) and (5) for the $n+1$ unknown functions u, E. In all cases, E will be of one sign; we take it to be positive, meaning the electric field is directed to the right.

There are two more quantities of considerable physical interest. The electric current J is defined to be

$$J \equiv z \cdot f, \tag{6}$$

and the electric potential ϕ is related to E by

$$E \equiv -\phi_x. \tag{7}$$

The various mathematical problems of evolution type arising in practice in electrophoresis are built by appending to these equations (4) and (5) various boundary and initial conditions, corresponding to the laboratory setup envisaged, as well as various assumptions regarding the parameters in the problem and the function g. In addition to these evolution problems, there are very important steady and traveling wave problems, as we will see.

Isotachophoresis (ITP). This is a particular type of electrophoresis in which separation occurs simultaneously with the development of a traveling wave. The species to be separated all have charges of the same sign (we will say positive for definiteness). There is also a negative ion to maintain approximate charge neutrality. The assumptions most commonly used in modeling the full isotachophoresis problem are:

(i) There are no reactions (the ions are fully dissociated; they cannot change from one into another).

(ii) The spatial domain is the entire real line.

(iii) Initially and forever afterwards, the mixture contains only one positive species u_{n-1}, with almost constant concentration, far to the right for x near $+\infty$ and it also contains only one positive species u_1, again with almost constant concentration, far to the left. Here u_{n-1} is called the *leading ion* (or *leading electrolyte*), and u_1 is the *terminating ion.*

(iv) u_1 has the least mobility (μ_1) of all the positive species, and u_{n-1} has the greatest.

(v) In addition to the positive species, there must be at least one negative ion u_n. We will assume that there is only one; it is called the *counterion.*

Being on an infinite domain with constant initial conditions near $\pm\infty$, the problem needs no boundary condition except for E. On the other hand, we can legitimately pose a traveling wave problem for ITP, in which case we have to be careful about the right choice of boundary conditions; we will discuss this further. A global existence theorem for the initial value problem in some important cases was proved by Avrin [Av]. The theory of traveling wave solutions for three species was studied by Fife, Palusinski, and Su [FPS].

Isoelectric focusing (IEF). Under proper conditions, the experimental observation in this case is that ampholyte molecules of each given type migrate to a specific location where they assume a final stationary concentration distribution. Separation is accomplished since the locations of these final distributions change from species to species. This is best modeled by a finite domain situation with boundary conditions imposed on the ends.

The terminology "ampholyte" refers to (usually protein) species that can exist in either positively, negatively, or neutrally charged conditions. For simplicity, we suppose that the charged ions can exist only with $z = \pm 1$ (the neutral state has $z = 0$, of course). For a given ampholyte A_i, we denote the three states by A_i^+, A_i^-, and A_i^0. The corresponding concentrations, mobilities, etc., have the same superscripts.

The reactions are

$$A_i^- + H^+ \rightleftarrows A_i^0, \tag{8a}$$

$$A_i^0 + H^+ \rightleftarrows A_i^+, \tag{8b}$$

where H^+ is the hydrogen ion. Let u_1 denote the concentration of H^+, and $u_i^\pm$, u_i^0 those of $A_i^\pm$ and A_i^0. Assume that both the forward and backward reaction rates are very high. Then these reactions exist in partial equilibrium: for some equilibrium constants K_i^0 and K_i^+,

$$u_i^- u_1 = K_i^0 u_i^0, \tag{9a}$$

$$u_i^0 u_1 = K_i^+ u_i^+. \tag{9b}$$

Suppose there are m ampholyte species, so $i = 1, \cdots, m$. Then we define m new variables by

(10) $$v_i \equiv u_i^- + u_i^0 + u_i^+,$$

the total concentration of that species irrespective of charge. We can use (9) to solve for all the u_i's in terms of v_i and u_1

(11) $$u_i^s = \rho_i^s v_i, \qquad s = -, 0, \text{ or } +,$$

where

(12) $$\rho_i^s = \frac{\sigma_i^s}{\sigma}, \quad \sigma_i^- = \frac{K_i^0}{u_1}, \quad \sigma_i^0 = 1, \quad \sigma_i^+ = \frac{u_1}{K_i^+},$$

$$\sigma = \sigma_i^- + \sigma_i^0 + \sigma_i^-.$$

In (4), the three charged varieties for a given i are distinguished as three different species. But we now group them as in (10). Consider the three equations in the system (4) corresponding to a given i. The corresponding three components of g add up to zero; this, in fact, results from the conservation of each ampholyte species A_i.

Now let us add those three equations and use (11) to get an equation for v_i

$$\partial_t v_i + \bar{\mu}_i(\bar{z}_i(v_i E)_x - \alpha \partial_{xx} v_i) = 0,$$

or again in m-vector form,

(13) $$v_t + \bar{\mu}(\bar{z}(vE)_x - \alpha v_{xx}) = 0,$$

where the barred vector constants $\bar{\mu}$ and $\bar{z}$ are simply averages,

(14) $$\bar{\mu} \equiv \Sigma \mu^s \rho^s, \qquad \bar{\mu}\bar{z} = \Sigma \mu^s z^s \rho^s,$$

the summations being over $s = -, 0$, and $+$. Notice that *$\bar{z}$ is not in general the average charge density in the sense of $\hat{z} \equiv \Sigma \rho^S z^S$*. Nevertheless the total charge density is

(15) $$\Sigma_{(i,s)} z_i^s \rho_i^s u_i^s = \Sigma_{(i)} \hat{z}_i v_i,$$

and Poisson's equation (5) becomes

(16) $$\beta E_x = \hat{z} \cdot v.$$

In addition to (13) and (16), for a complete problem there will be equations for u_1 and any other nonampholytic species present. The boundary conditions typically involve no-flux conditions for all the ampholytes, but Dirichlet or given flux conditions for u_1 and a negative nonampholyte.

2. Isotachophoresis.

2A. The transient problem. This has been studied in [BZY], but the most complete results have been obtained by Geng [Ge]. In both of these cases, the approximation is made in (4) and (5) that $\alpha = \beta = 0$. It is indeed true that β is very small, and although it appears to multiply a derivative in (5), it was shown in [FPS] that β may be safely set equal to zero in the traveling wave problem studied there. The assumption that $\alpha = 0$ says that the characteristic diffusion length is smaller than the characteristic size of the separated zones. This often seems to be a good approximation. With those approximations made, (4) and (5) become

$$u_t + \mu z(Eu)_x = 0, \qquad z \cdot u = 0. \tag{17}$$

Therefore, taking the scalar product of the left side with z, we obtain from (16)

$$0 = z \cdot u_t = -z \cdot (\mu z(Eu)_x) = -J_x. \tag{18}$$

Thus J is independent of x. The current, in fact, can be controlled as a function of t in experiments. For simplicity we take J to be constant. Then

$$J \equiv (\Sigma z_i^2 \mu_i u_i)E,$$

so

$$E = \frac{J}{\Sigma z_i^2 \mu_i u_i}, \tag{19}$$

and from (17),

$$\partial_t u + k \partial_x \left(\frac{u}{16 \Sigma z_i^2 \mu_i u_i} \right) = 0$$

for some vector k. We can redefine the dependent variables u so as to transform this equation to the form

$$\partial_t u + p \partial_x \left(\frac{u}{U} \right) = 0, \qquad U \equiv \Sigma u_i, \quad p \text{ a given vector.} \tag{20}$$

The general Riemann problem has been solved for the case of three positive species [Ge], and the results appear to be easily extendable to more species. Moreover, in that same work, the interaction of shocks was studied, and a global existence theorem was proved for initial data satisfying certain monotonicity conditions.

The Riemann problem corresponds to the case when the initial concentrations of all the ions are piecewise constant, an important and meaningful case. The resulting shocks proceeding from the initial discontinuities then may represent moving boundaries separating regions of pure electrolytes, so they and their interactions would describe the various stages of the separation process under that type of initial data.

The system (20) of conservation laws, although higher order, has some interesting special features. Consider the shock conditions when there are three equations, for example. Given a state $u^{(0)}$, the question of which new states $u^{(1)}$ can be reached from $u^{(0)}$ by a single shock can be answered as follows. Every such new state lies on one of three straight lines (in state space) passing through $u^{(0)}$. The velocities can be found in all cases. Moreover, if we want to connect the initial state to another state in the positive octant, this almost always can be done by a succession of three or fewer shocks, no rarefaction waves being needed. The shock connections appear to be the physically relevant ones.

2B. The traveling wave problem. This problem was formulated and a theory developed for it in the case of three species (two positive and one negative) by Fife et al. [FPS]. The parameters α and β were not taken to be zero. The significance of this particular problem is that in a fully developed isotachophoretic traveling wave, there will be moving thin boundaries between zones containing pure electrolytes (only one positive ion, say). This problem focuses on the structure of such a moving layer and examines the influence of diffusivity α and nonelectroneutrality β on the structure of that interface. To get a smooth transition it is necessary for α to be positive, but the solution for small β is uniformly close to the solution for $\beta = 0$ (the electrically neutral solution). The paper [FPS] contains a complete existence, uniqueness, and regularity theory for the problem. Along with this analysis, a feasible numerical method for the problem was developed [SPF].

3. Isoelectric focusing.

There are some cases in which the spatial distribution of the concentration u_1 of the hydrogen ion entering into the coefficients in (13) (i.e., the pH gradient of the solution) is known not to interact very much with the concentrations of the sample species. For example, it may be true that this distribution was effected by a special preparation of buffers together with electric current passing through the medium. Then the sample mixture of ampholytes (concentrations being denoted by the vector v in (13)) is introduced in small amounts, so as not to disturb the existing pH gradient. Su [Su] has obtained some important analytic and numerical results in this case.

In such a case, we may study (13) with $u_1(x)$ a given function. Recall that u_1 enters into the coefficients $\bar{\mu}$ and $\bar{z}$ by means of the quantities ρ_i^s in (12).

The flux $\bar{f}_i$ of the ith ampholyte is given by

$$\bar{f}_i = \bar{z}_i v_i E - \alpha \partial_x v_i. \tag{21}$$

First, let us consider the steady-state solutions of (14), under zero flux boundary conditions. These solutions are simply obtained by setting the flux equal to zero. Moreover, the fluxes of the various ampholytes decouple and they can be obtained separately. From (21) we have

$$\bar{z}_i v_i E = \alpha \partial_x v_i. \tag{22}$$

The solution of this differential equation (pretend we know $E(x)$) may have an interior maximum at the point where the derivative vanishes. That happens only where $\bar{z}_i = 0$.

This location will be the "isoelectric point" for that species i. Let us see how that can be found from knowledge of $u_1(x)$. From (14) we have, omitting the subscript i for simplicity,

$$\bar{z} = \frac{\Sigma \mu^s z^s \rho^s}{\Sigma \mu^s \rho^s} = \frac{\mu^+ \rho^+ - \mu^- \rho^-}{\bar{\mu}}.$$

This vanishes when the numerator does. From (12) we find the isoelectric point to be the position where

$$u_1(x) = \left(\frac{K^0 K^+ \mu^-}{\mu^+}\right)^{1/2} \tag{23}$$

Since we have left out the subscripts, this expression gives a different point for each species. They accumulate at different locations, thus effecting separation.

If $E = O(1)$, it follows from (22) and the fact that $\bar{z} = O(1)$ that the width of the concentration peak at each isoelectric point is $O(\alpha)$. For small α, therefore, the solution asymptotes to sharply defined peaks.

Conditions have been obtained by Su [Su] under which the solution of the initial value problem for (13) approaches the stationary solution just described.

CHAPTER 4

Waves in Excitable and Self-Oscillatory Media

1. One-dimensional problems.

1A. Introduction. Propagating interfaces similar to those considered in Chap. 1, §2, have proved to be crucial in modeling and explaining certain important chemical and biological wave phenomena ([OY], [CCL], [Fi76c], [TF], [Fi81], [Fi84a], [Fi84b], [Fi85], [KG81], [KG, ip], [KT], [ZK], [Zy84]). The best known examples of these waves are the propagation of signals along a nerve fiber and chemical waves in the Belousov–Zhabotinskii reagent ([Win72], [Win73], [Win74a], [Win78], [Win80], [Win87], [Zh], [Zy84], [Ty76], [TF]). However, in these applications the appropriate model does not usually consist merely of (24) in Chap. 1 in which the dependence of f on x represents a built-in nonuniformity in the properties of the medium. Rather, it very often consists of (24) in Chap. 1 with dependence on x replaced by dependence on the value of a second function $v(x, t)$, which evolves according to a separate nonlinear equation, with or without diffusion. Accordingly, we consider the system

$$\partial_t u = \epsilon \Delta u + \frac{1}{\epsilon} f(u, v), \tag{1a}$$

$$\partial_t v = D \Delta v + g(u, v), \tag{1b}$$

where D is the diffusivity of the second quantity v, and g is a smooth function whose properties will not be specified at this point. See [TF] for a justification of this model to describe the BZ reagent, for which $D = O(\epsilon)$. When $D = 0$, it includes the well-known and well-studied FitzHugh–Nagumo model for nerve impulse propagation.

The function f is the same as that in Chap. 1, §2, so in particular its two outer zeros are given by $u = h_\pm(v)$ satisfying (we assume for convenience) (25) in Chap. 1. The velocity of traveling wave solutions of (26) in Chap. 1 will now be denoted by $\bar{c}$ rather than $\bar{v}$. It depends on v because the function f does; the analogue of (27) in Chap. 1 is now

$$\bar{c} = Q(v). \tag{2}$$

The normal velocity of the interface, on the other hand, will be denoted by $c = c_0 = \epsilon c_1 + \cdots$.

1B. Interface dynamics with global functions $h(v)$ and $D = 0$. In the spirit of the preceding chapters, let us look for solutions of (1) with moving interfaces. Some of the arguments in this section are found in [Fi76b], [Fi84a], [Fi79c]. First, we consider the case when $D = 0$, and at the same time specialize to one space dimension.

The lowest order outer problem is

$$f(u_0, v_0) = 0, \tag{3a}$$

$$\partial_t v_0 = g(u_0, v_0). \tag{3b}$$

As the solution of (3a), we choose

$$u_0 = h_-(v_0), \quad x \in \mathcal{D}_-, \quad u_0 = h_+(v_0), \quad x \in \mathcal{D}_+. \tag{4}$$

Defining

$$G_\pm(v) \equiv g(h_\pm(v), v), \tag{5}$$

we obtain from (3b)

$$\partial_t v_0 = G_\pm(v_0), \qquad x \in \mathcal{D}_\pm. \tag{6}$$

Thus v_0 evolves according to an ordinary differential equation in the two outer regions.

The lowest order inner analysis is much the same as in Chap. 1, §2. The analogue of (32) in Chap. 1 is now

$$\partial_{\rho\rho} U_0 + c_0 \partial_\rho U_0 + f(U_0, V_0(\rho, s, t)) = 0, \tag{7a}$$

and we also have

$$c_0 \partial_\rho V_0 = 0. \tag{7b}$$

In this section we will not continue to higher order approximations, so for convenience omit the subscript "$_0$" in the following. Note that we are in one dimension, so the position of Γ may be denoted by $x = X(t)$. Here of course X is a function only of t, rather than of s and t as it was in Chap. 1, §2. The stretched coordinate across the interface is therefore defined as $\rho \equiv (x - X(t))/\epsilon$; this is the independent variable appearing in (7a).

When $c \neq 0$, we have from (7b) that $V(\rho, t) = V(t)$ is independent of ρ, and by the matching condition, this is equal to $v(X(t) \pm 0, t)$. It follows that under this condition, the outer function $v(x, t)$ is continuous across Γ. By the same argument as in (34) in Chap. 1, the lowest order velocity of the interface is

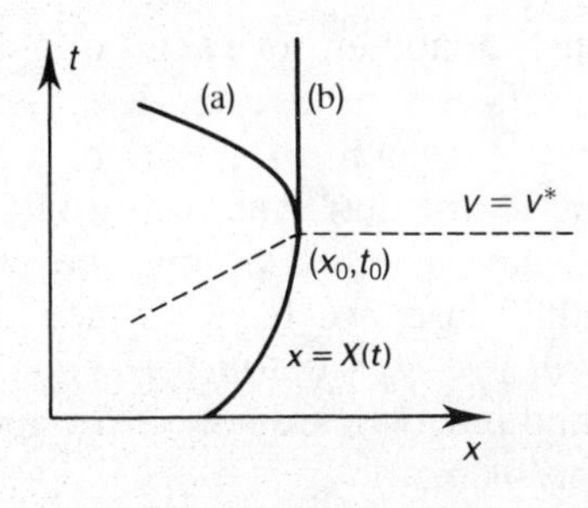

FIG. 4.1

given by

$$X'(t) = c = Q(v(X, t)). \tag{8}$$

Since the sharp wave fronts are in the variable u, and their velocities, by (8), are controlled by the value of v there, we sometimes call u the *propagator variable* and v the *controller variable* [Fi84a].

We have seen that v is continuous at points on Γ where $Q \neq 0$. This analysis needs to be supplemented by a discussion of what happens at places on Γ where $Q = 0$. This will be easiest if we impose a couple of additional hypotheses.

(9) $Q(v) = 0$ only at the single value $v = v^*$, where Q is an increasing function of v;

$$G_{\pm}(v^*) \neq 0. \tag{10}$$

At points where $Q = 0$, the outer function v is no longer necessarily continuous. But there is another condition that must be satisfied. To see it, let us suppose that Γ comes to rest first at some time t_0, so that

$$Q(v(X(t), t)) \neq 0, \quad 0 \leq t < t_0, \quad Q(v(X(t_0), t_0)) = 0.$$

For the sake of definiteness, assume $Q > 0$ for $t < t_0$. Then the trajectory $X(t)$ assumes the form shown in Fig. 4.1. The function v is continuous on Γ for $t < t_0$, and since its t-derivative is always bounded, it must also be continuous on Γ at the (rest) point where $t = t_0$. The curve $v(x, t) = v^*$ is also shown in Fig. 4.1; it is smooth except for a possible corner at the rest point.

This latter curve is as shown if $G_-(v^*) < 0$ (which we call case (a)), but if $G_-(v^*) > 0$ (case (b)), then there is no branch of that curve existing on the left of Γ.

In case (a), the function v will become less than v^* in the neighborhood of Γ for values of t slightly greater than t_0. In the lowest order layer equation (7), then, the function V appearing there will become less than v^*, which forces c to be negative. In view of (8), this means that Γ will reverse direction at that point, as shown. In other words, if $X'(0)$ is positive, then it is possible for $X'(t) = 0$ at only isolated values of t, and at those places, v continues to be continuous.

In case (b), when $G_-(v^*)>0$, then for values of t slightly greater than t_0, v is greater than v^* to the left of Γ in $\mathcal{D}_-$, and less than v^* to the right in $\mathcal{D}_+$. Clearly Γ cannot enter the region to the right where $Q=X'$ would be negative, nor the region to the left where it would be positive. The only possibility is for Γ to remain fixed. This may be termed a "locked front" phenomenon. But in either case, we have an inequality condition at points where $X'(t)=0$, namely $v(X-,t)\le 0\le v(X+,t)$.

The rules by which Γ and v evolve are now clear, and we restate them in the form of an initial value problem.

Reduced evolution problem P_1. Given a smooth initial function $v^0(x)$ and an initial point X^0 with $v^0(X^0)\ne v^*$, find functions $v(x,t)$ and $X(t)$, $t>0$, such that

(i) (6) holds with the interpretation $(x,t)\in\mathcal{D}_-$ for $x<X(t)$ and vice versa for $\mathcal{D}_+$;

(ii) for values of t with $X'(t)\ne 0$, v is continuous for all x, and

$$X'(t)=Q(v(X(t),t)); \tag{11}$$

(iii) for values of t with $X'(t)=0$,

$$v(X-,t)\le 0\le v(X+,t); \tag{12}$$

(iv) $X(0)=X^0$; $v(x,0)=v^0(x)$.

THEOREM 1. *Under conditions* (9), (10), *and the condition that the functions* $G_\pm(v)$ *be bounded,* P_1 *has a unique global solution* (v,X).

The existence part of the proof of this theorem proceeds by actual construction, which will be sketched. First, a provisional function v is constructed from (6) and (iv), assuming that the initial ($t=0$) separation of space into $\mathcal{D}_+$ and $\mathcal{D}_-$ persists for later times as well (which it does not). Then (11) is solved with the given initial value to obtain a provisional first front. This front is continued as long as $X'(t)\ne 0$. It serves to redefine $\mathcal{D}_+$ and $\mathcal{D}_-$ for all t up to that time. Then v is accordingly redefined behind the front. If $X'(t)$ vanishes at some point, this redefinition of v will determine whether the front becomes locked or doubles back. If it doubles back, the process continues.

The solution of P_1 can be used to provide the lowest order approximate solution of (1), (2). Namely, the function v represents the outer solution v_0; the outer solution u_0 is $h_\pm(v_0)$ in $\mathcal{D}_\pm$; the interface is given by $x=X(t)$ to lowest order; the lowest order inner solution V_0 is independent of ρ at places where $X'(t)\ne 0$ and equal to $v_0(X(t),t)$ there; finally the lowest order inner solution U_0, at those same places, is given by the solution of (7a) approaching the limits $u_0(X\pm,t)$ as $\rho\to\pm\infty$. This leaves only the inner functions at places where $X'=0$ unspecified. If v_0 is continuous at such a place, the inner functions are the same as those given above, but if it is not, then their determination requires a different sort of layer analysis which we will not go into here.

It is important to know the possible properties of solutions of P_1.

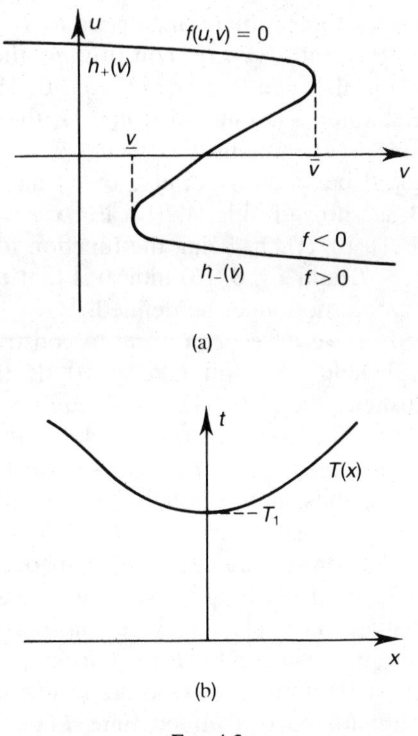

FIG. 4.2

THEOREM 2. *Assume* G_+ *and* G_- *are bounded away from zero,* $X^0=0$ *and* $Q(v^0(x))>0$ *for* $x \geq 0$. *Then under the conditions of Theorem* 1,

(a) *if* $G_+>0$, *then* $X(t)$ *is strictly increasing for all* t;

(b) *if* $G_+<0$ *and* $G_-<0$, *then either* $X(t)$ *is increasing for all* t, *or else there is a value* $t=t_1$ *such that* $X(t)$ *is strictly increasing for* $0 \leq t < t_1$, *and strictly decreasing for* $t_1<t<t_2$ *for some* $t_2>t_1$. *In the latter case if there is a second turning point (call it also* t_2*) at which* X *changes to become increasing again, then* $X(t_2)<0$;

(c) *if* $G_+<0$ *and* $G_->0$, *then either* $X(t)$ *is increasing for all* t, *or else there is a value* $t=t_1$ *such that* $X(t)$ *is strictly increasing for* $0 \leq t < t_1$, *and* $X(t) \equiv X(t_1)$ *for all* $t \geq t_1$ *(locked front).*

The cases when $Q(v^0(x))$ is not of one sign for positive x, or G_+ or G_- changes sign for some value of v, can be examined in a similar fashion.

1C. Nonglobal functions $h(v)$. In all of the applications of Theorem 2, the functions $h_\pm(v)$ are not globally defined, but rather come from a Z-shaped nullcline for the function $f(u, v)$ as shown in Fig. 4.2(a). Thus h_+ is only defined for $v \leq \bar{v}$, and h_- only for $v \geq \underline{v}$. In this case, the phenomena described above, of course, may continue to occur, but in addition new phenomena, such as the spontaneous appearance of interfaces, may arise.

To illustrate it, consider the problem analogous to P_1, except that initially there is no interface, the entire x-axis constituting the domain $\mathcal{D}_+$. Also suppose that $v^0(x) < \bar{v}$ for all x and that $G_+(v) \geq a > 0$. Then for each value of x, there will exist a first value $T(x)$ at which $v(x, t)$, the solution of (6) (with G_+) satisfying $v(x, 0) = v^0(x)$, attains the value $\bar{v}$. Suppose also that the function $T(x)$ (which will be smooth because v^0 is) has a minimal value T_1, attained at (say) $x = 0$, as shown in Fig. 4.2(b). For $x = 0$ and $t < T_1$, the outer function u_0 is simply $h_+(v_0(0, t))$, v_0 being the function v described previously in this paragraph. At $t = T_1$ and $x = 0$, (6) indicates that v_0 should continue to increase, but h_+ then would no longer be defined.

Therefore the procedure we have been using to construct an outer solution no longer will be applicable. We still take $x = 0$. If the requirement that $u_0 = h_+(v_0)$ is relinquished, then (1b) ($D = 0$) requires that v_0 continue to increase, forcing (u_0, v_0) into a region (to the right of the upper knee in Fig. 4.2) where $f(u, v) < 0$. In that region, the right side of (1a) will be large and negative (until $\partial_{xx} u_0$ becomes large enough), which in turn forces a large decrease in the value of u_0, again because of (1a). In so doing, the value of $u_0(0, t)$ will approach the lower part of the Z-shaped curve, i.e., the value $h_-(v_0(0, t)) \simeq h_-(\bar{v})$. This will occur quickly, so to a first approximation the time at which this transition takes place may be considered to be T_1.

This transition represents a change for fixed x from $\mathcal{D}_+$, where $u_0 \simeq h_+(v_0)$, to $\mathcal{D}_-$, where $u_0 \simeq h_-(v_0)$. If nothing else occurs to disturb the process, there will be such a transition for each x at the time $T(x)$. Therefore the curve $t = T(x)$ will be an interface separating $\mathcal{D}_+(t)$ from $\mathcal{D}_-(t)$. We call this a "downjump." (The analogous forced transitions from $\mathcal{D}_-$ to $\mathcal{D}_+$ when v reaches $\underline{v}$ are called "upjumps.") The location and movement of this front will depend on the initial datum $v^0(x)$.

Such interfaces, generated by v approaching its upper or lower limit $\bar{v}$ or $\underline{v}$, have been called *phase fronts,* and those described by the solution of P_1 have been called *trigger fronts.* (See [TF]; these terms were used by Winfree.) The phase fronts should be subject to the same kind of layer analysis as are the trigger fronts. Accordingly, at points on the phase interface, the lowest order inner layer analysis will produce (7) again. The differences now are as follows. First, by definition, the value of v on Γ is $\bar{v}$ or $\underline{v}$, so one of these values should take the place of V_0 in the second argument of the function f in (7a). Let us suppose it is $\bar{v}$. Second, the position and velocity of Γ is now known a priori; Γ is now simply the curve $t = T(x)$. Therefore the velocity c_0 in (7) is known and is equal to $(T'(x))^{-1}$. Again for definiteness, suppose this is positive, so T is an increasing function of x. The obvious question now arises as to whether there is a solution of (7) under these conditions, approaching the correct limits as $\rho \to \pm\infty$. We now observe that this equation is no longer of bistable type for $v = \bar{v}$, since the function $f(u, \bar{v})$ no longer has three distinct zeros. It has only two: $u = h_\pm(\bar{v})$, and the function is negative between those two zeros. Such equations are named after the geneticist R. A. Fisher, who used them in 1937 [Fis] to model genetic waves in spatially distributed populations. The

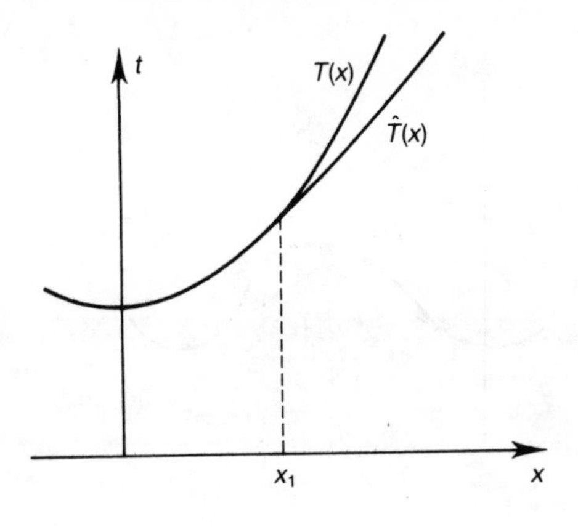

FIG. 4.3

difference between Fisher equations and bistable equations is that the former have traveling waves with a whole continuum of possible velocities (see, for example, [AW75], [Fi79c]). In this case, the possible velocities of fronts approaching $h_-(\bar{v})$ as $\rho \to -\infty$ and $h_+(\bar{v})$ at the other end satisfy the following inequality analogous to (8):

$$c_0 \geq Q(\bar{v}); \tag{13}$$

the quantity on the right here represents $\lim_{v \uparrow \bar{v}} Q(v)$.

The answer to the existence question posed above, then, is simply this: there is a solution of the inner layer equation (7) at points on the phase interface where

$$(T'(x))^{-1} \geq Q(\bar{v}). \tag{14}$$

For example, this will certainly be true at $x = 0$, where T has a minimum and the quantity on the left of (14) is infinite. However, there may be points on that interface at which (14) does not hold. Suppose there are such points for $x > 0$, and let x_1 be the largest value such that (14) holds for $x \in (0, x_1]$ (see Fig. 4.3). The phase front becomes a trigger front at exactly that value, $x = x_1$. The new trigger front, which we denote by $\hat{T}(x)$, is tangent to $T(x)$ at x_1. Because the slope of T is increasing at that point, $\hat{T}$ moves into a region to the right of the curve T, i.e., to where v has not yet attained the value $\bar{v}$. In that region, therefore, $v < \bar{v}$ and the trigger velocity $Q(v)$ is well defined.

We therefore have the phenomenon of a phase front being transformed into a trigger front. In solving the initial value problem stated earlier, the procedure would be to first find a *potential* phase interface T, disregarding any possible trigger interfaces developing. Then according to the criterion (14), find the points where they must become trigger fronts. Where the new fronts are below the potential phase fronts in the x,t plane, the latter should be

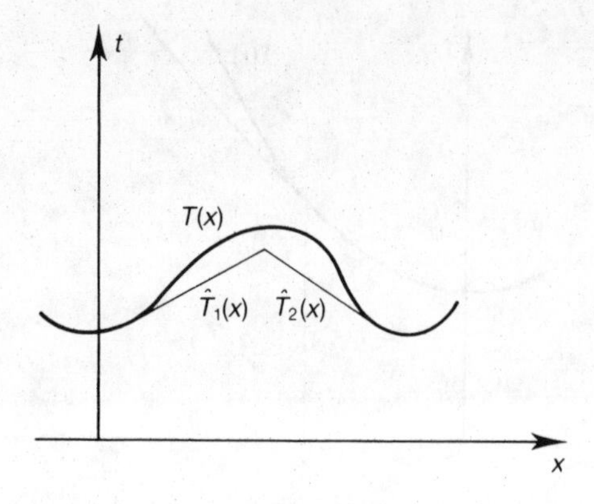

FIG. 4.4

deleted since they are not realized. However, it may happen that a trigger front produced in this way will intersect at some later time the potential phase front or another trigger front coming from the other direction. In that case, the two fronts annihilate each other (rather than passing through each other as in linear wave theory). This annihilation is illustrated in Fig. 4.4.

The case when the functions $h(v)$ are not global is therefore more multifaceted. It generates a reduced dynamical problem which is an extension of P_1. The rules will not be written down explicitly, but the important ones are implicit in the above discussion. A constructive procedure for solving an initial value problem for the more complicated system is as follows.

(1) Given initial data $v^0(x)$ together with initial positions of perhaps several fronts of either type, find potential "first" phase interfaces by (a) continuing in time those that are prescribed to begin at positions on the initial axis $\{t=0\}$, and (b) finding those that are generated spontaneously by v reaching its extremal value of $\bar{v}$ (in $\mathcal{D}_+$) or $\underline{v}$ (in $\mathcal{D}_-$).

(2) Find where such phase interfaces will change to trigger ones; then find the "first" trigger fronts by (a) continuing those that are prescribed to begin with, and (b) following those that are generated from the first potential phase fronts. If they intersect each other or a previously found potential phase front, they both terminate at that point. Follow these trigger fronts until they intersect something, become locked, or turn back.

(3) Find the next potential phase front above (later than) the interfaces found in (1) and (2) above.

(4) Find the next trigger fronts by continuing those that were locked in step (2) or turn back, and by finding those that are generated from the next phase fronts. Terminate them when they intersect. Continue this procedure.

1D. The case $D=\alpha\epsilon$. This case, when α is an $O(1)$ parameter, is important in modeling the Belousov–Zhabotinskii waves. It represents the case when the two chemicals u and v have diffusivities that are the same order of magnitude. This case can be handled in the same way as the case $D = 0$, and the results are in fact the same, to lowest order.

The lowest order outer equations (3) remain the same; again we choose u_0 and v_0 to satisfy (4) and (6). The lowest order inner equation for U_0 also remains the same; the first difference is in the inner equation for V_0, which is now

$$c_0 \partial_\rho V_0 + \alpha \partial_{\rho\rho} V_0(\rho, t) = 0. \tag{15}$$

This and the boundedness of V_0 imply that V_0 is independent of ρ, so that by lowest order matching, the outer function v_0 is continuous. This same conclusion was reached before, in the case $D = 0$, at points of Γ where the velocity $X' \neq 0$, and isolated points where $X' = 0$. The remaining case is that of a locked interface; it was mentioned for this case that (i) v_0 can be expected to be forced to be discontinuous, and (ii) a different layer asymptotics is appropriate. In fact, both of these conclusions are also true in the present situation. The argument in this paragraph above that v_0 is continuous breaks down at locked fronts, because in general the continuity of v_0 there is incompatible with the dynamics (6) when X is constant. Equation (15) is not valid for locked fronts; in fact, ρ is an inappropriate inner variable there. Again, a different layer asymptotics is appropriate. The details turn out to be unlike those for the case $D = 0$.

2. Inhomogeneous media.

Interfacial waves of the type described in §1 but which propagate in a medium with spatially varying properties may be modeled by allowing the functions f and g in (1) to depend on x as well as on u and v. The concepts and techniques differ little from those in the preceding section. The x-dependence is carried through to the functions Q, $h_\pm$, and $G_\pm$ in (2), (4), (5), etc. The trigger wave propagation law (11) becomes

$$X'(t) = Q(v(X(t), t), X(t)). \tag{16}$$

In the case of nonglobal functions $h_\pm$, the limits $\bar{v}$ and $\underline{v}$ will also depend on x. The discussion following (14) will remain the same.

Of special interest is the case when the medium is self-oscillatory in at least one region of space, and the natural frequency of oscillation is not constant. We need to explain the meaning of this term. Let $\phi_+(x)$ denote the time it takes the function v, satisfying (6_+) (the subscript "$_0$" omitted) to evolve from $\underline{v}(x)$ to $\bar{v}(x)$, if it does so at all. If it cannot so evolve, we say that $\phi_+(x) = \infty$. Similarly let $\phi_-(x)$ denote the time of passage from $\bar{v}(x)$ to $\underline{v}(x)$ under the action of (6_-). Then the natural period at any point x, to lowest order, is $\phi(x) \equiv \phi_+(x) + \phi_-(x)$. We define the medium to be "self-oscillatory" at the point x if $\phi(x)$ is finite.

It is generally said to be "excitable," on the other hand, if equation (6), with either $+$ or $-$, but not both (let us say "$-$" for definiteness), has an attracting

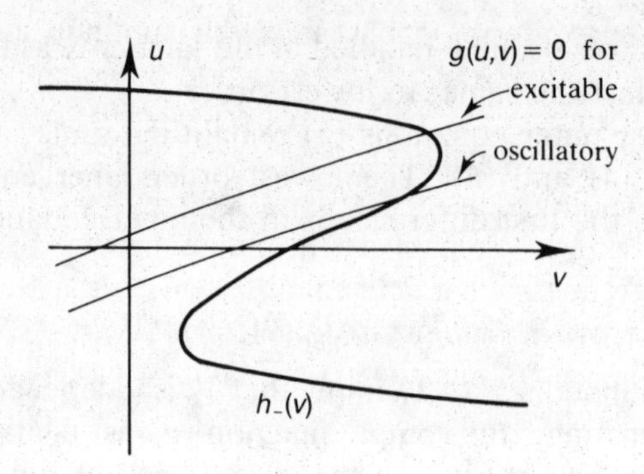

FIG. 4.5

rest state $\bar{v}$ that is near $\underline{v}$ and $\phi_+ < \infty$. An oscillatory state can be converted into an excitable one by a slight change of the nullcline $g = 0$ (see Fig. 4.5).

We consider an initial value problem for a self-oscillatory medium, and for the sake of illustration consider the initial datum to be constant, and the natural frequency to be maximal at $x = 0$. Assume the initial state is $\mathcal{D}_+$ for all x. For convenience let us take $v_0(x) \equiv \underline{v}$. Then the first phase front (a downjump), resulting from the arrival of v at the value $\bar{v}$, occurs at the time $T_{1+}(x) \equiv \phi_+(x)$, and the next one (an upjump) occurs at $T_{1-}(x) \equiv \phi(x)$. Continuing this way, we find the nth potential downjump phase front occurs at time

$$T_{n+}(x) \equiv n\phi(x) - \phi_-(x). \tag{17}$$

Our assumption is that $\phi(x)$ has a strict minimum at $x = 0$. We ask now about the possibility of trigger fronts developing. According to (14), downjump phase fronts will generate them when the criterion (14) is first violated, i.e., when $T'_{n+}(x)$ surpasses the critical value $(Q(\bar{v}, x))^{-1}$. Since $\phi'(x) > 0$ for $x > 0$ small, we see by differentiating (17) that this criterion will indeed be satisfied for large enough n, for some small positive $x = x_n$. In fact, as n increases, this transition point grows closer to the origin:

$$\lim_{n\to\infty} x_n = 0. \tag{18}$$

For the same reason, left-moving trigger fronts develop to the left of the origin.

We conclude that the appearance of trigger fronts is inevitable when the natural frequency has a maximum. The subsequent motion of these trigger fronts depends, of course, on the properties of the medium. Suppose, for example, that it is homogeneous outside some interval containing the origin: $\phi(x) \equiv \bar{\phi}$ for $|x| > x_0$. If $\bar{\phi}$ is finite (so the medium is self-oscillatory for all x), then necessarily all the trigger fronts generated near the origin will intersect phase fronts, possibly in the region $|x| > x_0$ where the latter are horizontal lines in the $x - t$ plane, and thus disappear.

On the other hand, if $\bar{\phi} = \infty$, then no such intersection occurs in that region and any trigger front entering the region will continue indefinitely. In that case, as $t \to \infty$ the function $v(x, t)$ in the region $x > x_0$ approaches a periodic traveling wave moving to the right with period $\phi(0)$. Moreover, the fronts' velocities approach a constant.

The analogous phenomenon in two space dimensions represents the periodic appearance of expanding circular fronts that are generated near the origin, where the natural frequency has a maximum, and propagate into the region of space where it is zero. These expanding target patterns are commonly seen in the BZ reagent. The above mechanism for their generation was proposed and studied in detail by Tyson and Fife [TF].

3. Curvature, corners, and twists.

3A. Corrections to the velocity due to curvature and dependence on v. We consider a solution of (1) in the plane with an interface $\Gamma(t)$. If $D = 0$, the lowest order outer solution is governed still by (4) and (6). For any fixed time t and for any point X on Γ at which the normal velocity c is nonzero, the lowest order inner problem will consist of (7); hence

$$\partial_\rho V_0 = 0, \tag{19}$$

so that $V_0 = V_0(s, t)$ is independent of ρ. According to the matching condition (5a) in Chap. 1, V_0 is equal to the limit of the outer function v_0 as Γ is approached from either side. We conclude, as before, that v_0 is continuous across Γ at points where its normal velocity is not zero, and that $V_0 = v_0(X(s, t), t)$.

As in (35) in Chap. 1, the solution of (7) is

$$U_0(\rho, s, t) = \psi(\rho, v_0(X(s, t), t)), \tag{20}$$

where X is the position on Γ with coordinates (s, t).

The lowest order normal velocity is given, as in (8), by

$$c_0(X, t) = Q(v_0(X, t)). \tag{21}$$

The next order inner problem consists of the following equations:

$$\begin{aligned} \partial_{\rho\rho} U_1 + c_0 \partial_\rho U_1 + f_u(U_0, v_0(X(s, t), t))U_1 \\ = (-c_1 - \kappa(s, t))\partial_\rho \psi(\rho, v_0(X, t)) - f_v(U_0, v_0)V_1 \\ + (\partial_t + (\partial_t s_0)\partial_s)\psi(\rho, v_0(X(s, t), t)), \end{aligned} \tag{22}$$

which is analogous to (37) in Chap. 1, and

$$-c_0 \partial_\rho V_1 = -(\partial_t + (\partial_t s_0)\partial_s)v_0(X(s, t), t) + g(U_0(\rho, s, t), v_0(X(s, t), t)). \tag{23}$$

The operator appearing in the last term of (22) and the first term on the right of (23) is the operator $\hat{D}$ discussed in (38) and (41) in Chap. 1. We would like to apply (41) in Chap. 1 to the function $F(\rho, x, t) \equiv v_0(x, t)$. However, the latter function is not smooth; it has discontinuous derivatives on Γ. Instead, let

$v_0^+(x, t)$ be a smooth function with $v_0^+(x, t) \equiv v_0(x, t)$ for $x \in \mathcal{D}_+(t)$. In other words, it is a smooth extension of the restriction of v_0 to $\mathcal{D}_+$. Similarly, let v_0^- be smooth and identical to v_0 in $\mathcal{D}_-(t)$.

We may apply (41) in Chap. 1 (replacing v by c, of course) to either v_0^+ or v_0^- to obtain, to lowest order,

$$\hat{D}v_0^\pm(X(s, t), t) = \partial_2 v_0^\pm(X(s, t), t) + c_0 \partial_r v_0^\pm(x, t)_\Gamma. \tag{24}$$

By (6), the first term on the right is $G_\pm(v_0(X, t))$, where the superscript "$\pm$" on v_0 is now omitted, because v_0^+ and v_0^- coincide on Γ. The same argument can be used to transform the last term in (22). As a result, we have, in place of (22) and (23),

$$\begin{aligned}\partial_{\rho\rho} U_1 + c_0 \partial_\rho U_1 + f_u(U_0, v_0(X(s, t), t))U_1 \\ = (-c_1 - \kappa(s, t))\partial_\rho \psi(\rho, v_0) - f_v(U_0, v_0)V_1 \\ + \psi_v(\rho, v_0(X, t))(G_\pm(v_0(X, t)) + c_0 \partial_r v_0^\pm(x, t)|_X),\end{aligned} \tag{25}$$

$$-c_0 \partial_\rho V_1 = -c_0 \partial_r v_0^\pm(x, t)|_X - G_\pm(v_0(X, t)) + g(U_0(\rho, s, t), v_0(X(s, t), t)). \tag{26}$$

In (25) and (26), either sign, "+" or "−," may be chosen; they give the same result.

Applying an orthogonality condition to (25) now results in the following relation, analogous to (43) in Chap. 1:

$$(-c_1 - \kappa)A(s, t) + \hat{B}(s, t) = 0, \tag{27}$$

where A is the same as in (43) in Chap. 1, and

$$\hat{B} = \int p^*(\rho)[-f_v(U_0, v_0)V_1 + \psi_v(\rho, v_0(X, t))(G_\pm(v_0(X, t)) + c_0 \partial_r v_0^\pm(x, t)|_X)]\, d\rho.$$

Combining (21) and (27), we have that the interface velocity, to order $O(\epsilon)$, is given by

$$c(X, t) = Q(v_0(X, t)) - \epsilon\kappa(X, t) + \epsilon \hat{B}(s, t)/A(s, t). \tag{28}$$

The $O(\epsilon)$ correction to the velocity, therefore, consists of two terms: one is proportional to the curvature of Γ and the other results from the dependence of f on v.

The effect of curvature on the velocity of a wave front has been examined by Zykov, Keener, and Tyson ([Zy80a], [Zy80b], [Zy84], [Ke86], [KT]), who use a different approach to obtain an equation like (28), but with the last term missing. They show, in fact, that for steady or steadily rotating solutions, the relation

$$c(X, t) = Q(v_0(X, t)) - \epsilon\kappa(X, t) + O(\epsilon)$$

holds even without the restriction $\kappa = O(1)$ which is implicit in our asymptotic development. They do this by using a local coordinate system (ρ', s) such that ρ' is constant on each level curve of the function U. The context they have in mind is when $\kappa > O(1)$ and the middle term in (28) is of a larger order of magnitude than the last, so the last one may be negligible.

The inner function $V_1(\rho, s, t)$ is obtained by integrating (26). The matching relation (5b) in Chap. 1 is checked by letting $\rho \to \pm\infty$ on the right. In the case of $+\infty$, we choose the "+" sign in (26); the difference of the last two terms then approaches 0. This is similar to when $\rho \to -\infty$. We thus obtain, automatically, the matching relation

$$\partial_\rho V_1|_{\rho=\pm\infty} = \partial_r v_0(X \pm 0, t),$$

which conforms with (5b) in Chap. 1. Integrating (26) introduces an integration constant $C(s, t)$, which should be chosen in order to satisfy the undifferentiated part of the matching condition (5b) in Chap. 1. This latter condition requires independent knowledge of the next order outer solution $v_1(X, t)$. But v_1 can be obtained by integrating the next order outer equation ((6) is the lowest order), so it depends on the initial data. We will not dwell on this point.

We will be examining two cases in which the curvature correction to the velocity is no longer relatively small, but rather comparable in magnitude to the first term on the right of (28). The first such case involves a corner layer, in which both terms are $O(1)$, so that $\kappa = O(\epsilon^{-1})$. The second case is the formation of a vortex by a twist action, in which the two terms have equal orders of magnitude smaller than $O(1)$.

3B. Corner layers. Considering one-dimensional problems in §1, we found that a head-on collision of two wave fronts destroys both of them. In two dimensions, they may collide at an oblique angle rather than head on. In such a case, only the colliding parts of the fronts are annihilated.

If the original fronts are straight and move at a constant velocity, for example (as seen in Fig. 4.6), we then expect that the collision process will form an angular domain that moves at a constant velocity in the direction halfway between the directions of the two planar fronts. Moreover, the velocity c of this corner will be larger than the normal velocity c^* of the planar fronts, by the factor csc θ:

$$c = c^* \csc \theta. \tag{29}$$

We assume that $c^* > 0$ and that the domains $\mathscr{D}_\pm$ are as shown. The overall intuitive picture described is obtained when we neglect any consideration of the internal structure of the interface. Of course, it cannot be accurate very near the corner because it predicts an infinite curvature for Γ there, which would not be allowed in a solution of the basic partial differential equations (1). The corner actually must be smoothed out and have a finite curvature. (See Fig. 4.7.)

Assuming that the outer function v_0 is constant along Γ, we see from (29) that the normal velocity c at the corner differs from $c^* = Q(v_0)$ by an $O(1)$ amount, so the correction is not small and the above perturbative analysis is

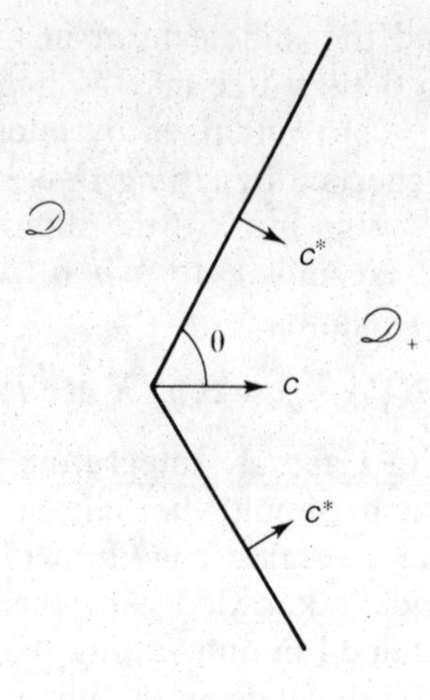

FIG. 4.6

not necessarily valid. Nevertheless, the form of the curvature correction (28) suggests that at the corner, $\kappa = O(\epsilon^{-1})$. If this is true, then in that neighborhood the structure of the interface has characteristic spatial dimensions $O(\epsilon)$ in all directions, rather than just in the direction normal to Γ.

Accordingly, to examine that structure we stretch both the x and the y coordinates by the factor ϵ^{-1}. For simplicity we do the analysis assuming a steady traveling wave configuration as described above, with $v_0 = \text{const}$ on the entire interface. Then (29) holds. We also take $D = 0$. Both of these assumptions may be relaxed or discarded by doing a more detailed analysis.

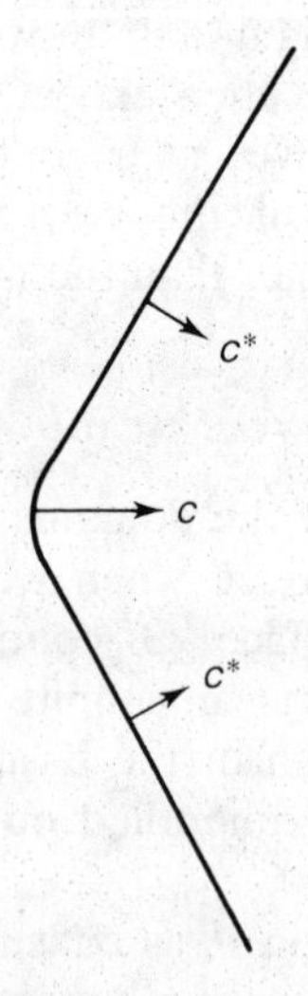

FIG. 4.7

The traveling wave form of (1) is

(30a) $$\epsilon \Delta u + c\partial_x u + \frac{1}{\epsilon} f(u, v) = 0,$$

(30b) $$c\partial_x v + g(u, v) = 0.$$

In this traveling coordinate system, the origin is always located at the corner. We rescale

$$\xi = x/\epsilon, \quad \eta = y/\epsilon, \quad U(\xi, \eta) = u(x, y), \quad V(\xi, \eta) = v(x, y),$$

and so obtain from (30)

(31a) $$\Delta_{(\xi,\eta)} U + c\partial_\xi U + f(U, V) = 0,$$

(31b) $$c\partial_\xi V + \epsilon g(U, V) = 0.$$

This last equation says that to lowest order in the corner layer, V does not depend on ξ. The inner variable V is therefore constant across Γ at the corner. We already knew this to be true away from the corner as well. Thus $V = v_0 = \text{const}$ on Γ. For convenience, the subscript "$_0$," when referring to the lowest order terms, will be omitted. We are left with the problem of finding a solution of (31a), with given fixed V, that matches the interfacial structure away from the corner layer.

The lowest order inner solution away from the corner is easily obtained; it consists of planar interfaces at an angle θ with the ξ-axis, moving in the directions indicated in Fig. 4.7. If ρ is the stretched normal variable, the inner profile $U^*(\rho)$ satisfies

(32) $$\partial_{\rho\rho} U^* + c^* \partial_\rho U^* + f(U^*, V) = 0,$$

whose solution is

(33) $$U^*(\rho) = \psi(\rho, V).$$

For the plane wave on the top, $\rho = \xi \sin\theta - \eta \cos\theta$; on the bottom it is $\rho = \xi \sin\theta + \eta \cos\theta$. In the following discussion, dependence on V will not be shown explicitly; for example, we write $f(U)$ instead of $f(U, V)$.

Our problem is now seen to consist of finding a solution $U(\xi, \eta)$ of (31a) such that

(34a) $$|U(\xi, \eta) - \psi(\xi \sin\theta - \eta \cos\theta)| \to 0 \quad \text{as } (\xi, \eta) \to \infty, \quad \eta > 0,$$

(34b) $$|U(\xi, \eta) - \psi(\xi \sin\theta + \eta \cos\theta)| \to 0 \quad \text{as } (\xi, \eta) \to \infty, \quad \eta < 0.$$

Such a solution cannot be given explicitly, of course, but we can prove that it exists under a certain condition. The method involves constructing an upper and a lower solution of (31a) in all (ξ, η) space, both of them satisfying (34). Then by a well-known theorem (an account is given in [Fi79c], for instance) there must exist an exact solution between them.

Let the nonlinear differential operator N be defined by

$$NU \equiv -cU_\xi - \Delta_{\xi\eta} U - f(U),$$

so that (31a) is simply

$$NU = 0. \tag{35}$$

Upper solutions are functions $\bar{U}$ satisfying $N\bar{U} \geq 0$ (or minima of sets of such functions); lower solutions $\underline{U}$ satisfy the opposite inequality. It is very easy to find an upper solution satisfying (34): we merely take

$$\bar{U}(\xi, \eta) \equiv \text{Min}[\psi(\xi \sin\theta - \eta\cos\theta), \psi(\xi\sin\theta + \eta\cos\theta)]. \tag{36}$$

The lower solution is more difficult. We look for one in the form

$$\underline{U}(\xi, \eta) = \psi(a(\xi) - b(\eta)) \tag{37}$$

for some appropriate choice of functions a and b that satisfies

$$\lim_{\xi\to\infty} (a(\xi) - \xi\sin\theta) = \lim_{\eta\to\pm\infty} (b(\eta) \mp \eta\cos\theta) = 0. \tag{38}$$

Using (29) and the fact that $\psi(\rho)$ satisfies (32), we set $\rho(\xi, \eta) \equiv a(\xi) - b(\eta)$ and calculate

$$\begin{aligned} N\underline{U}(\xi, \eta) = & -c^* \csc\theta a'(\xi)\psi'(\rho) - (a'(\xi))^2\psi''(\rho) - a''(\xi)\psi'(\rho) \\ & - (b'(\eta))^2\psi''(\rho) + b''(\eta)\psi'(\rho) - f(\psi) \\ = & [(a')^2 + (b')^2 - 1](c^*\psi' + f(\psi)) \\ & + [c^*(a'^2 + b'^2) - c^* \csc\theta a' - a'' + b'']\psi'. \end{aligned} \tag{39}$$

Since $f'(\psi) \neq 0$ at the two outer zeros $\psi = h_\pm(V)$ of f, it is seen that $\psi(\rho)$ and $\psi'(\rho)$ approach their limits as $\rho \to \pm\infty$ exponentially at the same exponential rate, and that the function $f(\psi(\rho))$ decays at this same exponential rate as $\rho \to \pm\infty$ as well. These facts, together with the fact that $\psi'(\rho) > 0$, imply that for some positive constant α,

$$|f(\psi(\rho))| \leq \alpha\psi'(\rho) \quad \text{for all } \rho. \tag{40}$$

Substituting this estimate into (39), we obtain

$$N\underline{U} \leq (A + B)\psi'(\rho), \tag{41}$$

where

$$A = A[a] = (a'(\xi))^2(\alpha + c^*) - a''(\xi) - c^* \csc\theta a'(\xi) - \alpha_1, \tag{42a}$$

$$B = B[b] = (b'(\eta))^2(\alpha + c^*) + b''(\eta) - \alpha_2, \tag{42b}$$

and where α_1 and α_2 are any constants that add up to α.

We proceed by finding functions a and b to satisfy $A[a] = B[b] = 0$ and (38). In order for (38) to be possible, we choose

$$\alpha_1 = \alpha\sin^2\theta - c^*\cos^2\theta, \qquad \alpha_2 = (\alpha + c^*)\cos^2\theta.$$

It may be checked that the following function b satisfies these conditions:

(43) $$b(\eta)=\int_0^\eta q(s)ds \quad \text{where } q'(\eta)=-(\alpha+c^*)(q^2-\cos^2\theta),$$
$$q(\pm\infty)=\pm\cos\theta.$$

As for a, we set

(44a) $$a(\xi)=\int_0^\xi p(s)\,ds+a_0,$$

where a_0 is for the moment arbitrary, and

(44b) $$p'(\xi)=m(p)\equiv(\alpha+c^*)p^2-c^*\csc\theta\, p-\alpha\sin^2\theta+c^*\cos^2\theta.$$

It may be checked directly that $m(\sin\theta)=0$ and $m'(\sin\theta)>0$, provided that

(45) $$2(\alpha+c^*)\sin^2\theta>c^*.$$

In this case, $m(p)$ has exactly one other zero, and that is less than $\sin\theta$. By checking the sign of $m(0)$, we see that the other zero p_1 is also positive, provided that

(46) $$c^*>\alpha\tan^2\theta.$$

We want both (45) and (46) to hold. However, note from the definition of α in (40) that the constant α may be chosen as any number larger than the minimum value α_0 for which (40) holds. In view of this, we consider two cases:

(i) $2\sin^2\theta\geq 1$. Then (45) holds automatically, and poses no restriction.

(ii) $2\sin^2\theta<1$. Then (45) is equivalent to $c^*\leq\alpha S(\theta)$, where

$$S(\theta)=\frac{2\sin^2\theta}{1-2\sin^2\theta}>\tan^2\theta.$$

Thus (45) and (46) can be expressed as $\alpha\tan^2\theta<c^*<\alpha S(\theta)$. Clearly, $\alpha\geq\alpha_0$ may be chosen so this is true, provided only that

(47) $$c^*>\alpha_0\tan^2\theta.$$

Suppose, then, that (47) holds. Then α may be chosen so that (45) and (46) also hold, which means that the two roots of $m(p)$ are positive. Hence, there is a bounded solution $p(\xi)$ of (44b) satisfying $p(\xi)>p_1>0$ for all ξ, and $\lim_{\xi\to\infty}p(\xi)=\sin\theta$. It is not unique since its argument may be translated at will; fix such a solution and denote it by $p(\xi)$.

Now defining $a(\xi)$ by (44a), we see that $a(\xi)$ is strictly monotone. Let $\Xi(a)=\xi$ denote its inverse $a(\Xi(a))\equiv a$.

Consider now the curve Γ in the (ξ,η) plane defined by

(48) $$\Gamma:\quad a(\xi)=b(\eta).$$

It can be rewritten in the form $\xi=\Xi(b(\eta))$, so it exists and is unique. Moreover, it is asymptotic to the lines $\xi=(p(\infty))^{-1}q(\pm\infty)\eta+\xi_1=\pm\eta\cot\theta$

as $\eta \to \pm\infty$, where the number ξ_1 depends on a_0. Specifically, $\xi_1 = \csc\theta[b_1 - a_1 - a_0]$, where

$$a_1 = \int_0^\infty [p(s) - \sin\theta]\,ds, \qquad b_1 = \int_0^\infty [q(s) - \cos\theta]\,ds.$$

We choose $a_0 = b_1 - a_1$, so that $\xi_1 = 0$. Then (38) will hold.

This line Γ is, according to (37), the line on which $\underset{\sim}{U} = 0$ (since we have defined ψ in (35) in Chap. 1 so that $\psi(0) = 0$). If $\underset{\sim}{U}$ were the exact solution, Γ would be the real interface if we define that as the line where $\underset{\sim}{U} = 0$.

This completes the definition of $\underset{\sim}{U}$. As mentioned, there is an exact solution $U(\xi, \eta)$ satisfying $\underset{\sim}{U}(\xi, \eta) \le U(\xi, \eta) \le \bar{U}(\xi, \eta)$, and it satisfies (34) as well.

In summary, if θ is small enough so that (47) holds, then there exists a solution of the corner layer problem (31a), (34). A solution in fact probably exists (conjecture) with that restriction on θ removed.

3C. The twisting action. There is a very important idiosyncrasy that solutions of (1) have in two (or more) space dimensions. Under well-defined conditions (which will be evident from the arguments below), solutions with regular initial data immediately will develop singularities of a sort reminiscent of hydrodynamic vortices in shear flow with an interface in the form of a discontinuity in the tangential velocity. The similarity is, of course, only superficial. In the present case, it is not a true singularity that develops; rather, the curvature of the interface becomes so large near certain points that the lowest order approximate solution method described in §3A is not valid.

This peculiarity is best illustrated by the following simple initial value problem. Let (x, y) be the spatial coordinates. Assume (9) and (10) hold, and also $G_+(v) > 0$, $G_-(v) < 0$. We assume that at $t = 0$, the interface coincides with the x-axis: $\Gamma(0) = \{y = 0\}$. Moreover, $v(x, y, 0) \equiv v^0(x, y) = v^* + \phi(x)$, where v^* is the constant in (9) and ϕ is an odd strictly increasing function like $\tanh^{-1} x$ which vanishes at $x = 0$ and approaches limits at $\pm\infty$. Most of these assumptions are for simplicity only; the phenomenon to be described is certainly generic.

According to (8) and (9), the normal velocity $c(X)$ of $\Gamma(0)$ at the point $X = (x, 0)$ is a function of x similar in nature to ϕ. If the position of Γ could be approximated a short time Δt later by naively extending it a distance $c(X)\Delta t$ in the vertical direction at each point X, then it would assume the form indicated in Fig. 4.8. During this small time interval, the function v in the domain $\mathcal{D}_+$ would have evolved according to (6), and in particular at the point $A \in \Gamma(\Delta t)$ near the origin, since $G_+ > 0$, v will be approximated by $v^* + O(\Delta t)$. Similarly, at B, $v \simeq v^* - O(\Delta t)$. This means that if (8) were still operative, then the new normal velocity would be positive and $O(\Delta t)$ at A; it would be the same but negative at B. Since these points are arbitrarily close to O and Δt is fixed, the angular velocity of the twisting of Γ at O would be infinite, and hence in that short time interval Γ would have rotated an infinite number of times, contrary

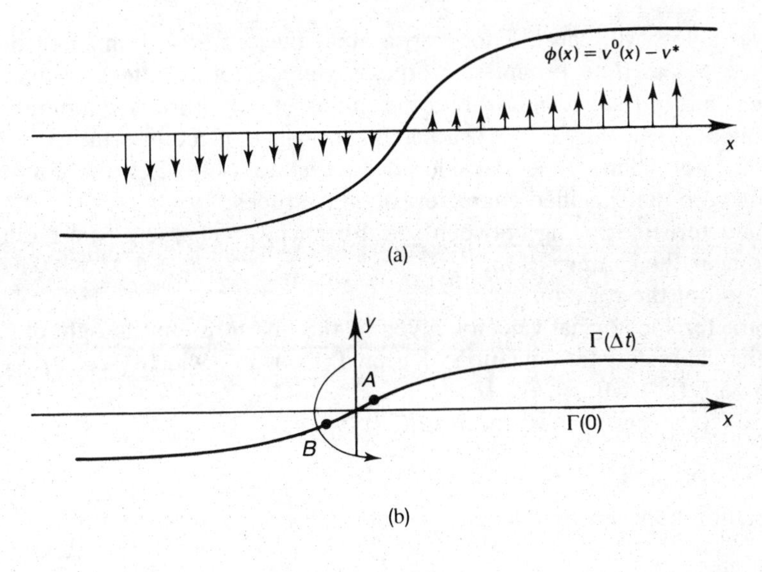

FIG. 4.8

to Fig. 4.8. Its shape instead would be extremely singular at the origin, with points of curvature κ arbitrarily large.

The solutions of (1) do not behave this way, of course; the fact is that the conditions for validity of our asymptotics immediately break down in a neighborhood of the origin. A clue to what actually happens can be taken from (28), which shows $\epsilon\kappa$ is no longer a small correction. Using the lowest order formal approximation (21) led to nonsense. The normal velocity must be affected by the curvature of Γ in a manner akin to (28); large curvature tends to reduce the interfacial velocity.

It would be desirable to set up some alternate type of asymptotic procedure that would provide an adequate description of the interface dynamics in a neighborhood of the twist point. We will pursue this matter only so far as to suggest a rescaling of space and time, which should do the job, and (in the next section) to formulate a problem for the steady form of the solution after a long period of time has elapsed.

We will argue that the rescaling should be

$$x = \epsilon^{2/3}\xi, \quad t = \epsilon^{1/3}\tau. \tag{49}$$

The curvature of Γ is initially the principal effect that keeps the angular velocity finite. Its mitigating effect is through the second term on the right of (28). At the times and places when this term plays a decisive role in the interfacial dynamics, it is reasonable to suppose that the first two terms on the right of (28) (the first providing the propagation rule when there is no singularity, and the second providing the principal mitigating influence) have the same orders of magnitude. It is also reasonable to think that even though the singularity will be regularized by the effects of curvature, the tendency to twist nevertheless will cause Γ to wrap around the center a number of times, which increases as time progresses. Finally, this curvature effect occurs near the center, and in that neighborhood the radius of curvature is comparable to the spacing between the first few successive branches of the interface.

We will show that these considerations dictate (49). Let the following symbols denote the specified characteristic magnitudes:

L = characteristic spacing between spiral branches = characteristic radius of curvature near the center;

T = period of the rotation;

c = characteristic normal velocity, given by the first term on the right of (28);

Δv = the characteristic deviation of v on Γ from v^*, which (see (9)) is the value where $Q(v^*)=0$.

Then equating the first two terms in (28), we get

$$c \simeq \epsilon L^{-1}.$$

The relation between wavelength, period, and normal velocity is

$$cT \simeq L.$$

Assuming $Q' \neq 0$, then

$$c \simeq \Delta v.$$

Finally, for a fixed position in space, the time between the moment an upjump front passes that point and the moment when the next downjump passes is comparable to T, as is the time from that second moment to the next upjump. Between these jumps, the outer function v_0 is governed by ($6_\pm$). At the jump locations themselves, v_0 is alternately greater than v^* and less than v^*, because when viewed always from $\mathscr{D}_+$, the velocity of the interface is alternately positive and negative. Assuming $G = O(1)$, we therefore obtain from (6)

$$\Delta v \simeq T.$$

Combining these four relations easily yields

$$L \simeq \epsilon^{2/3}, \qquad T \simeq c \simeq \Delta v \simeq \epsilon^{1/3}, \tag{50}$$

hence (49).

4. Rotating waves.

4A. The questions. The corner layer represents a case of a propagating curved interface for which the first two terms on the right of (28) are each $O(1)$; in particular, $\kappa = O(\epsilon^{-1})$, so the radius of curvature equals $O(\epsilon)$. This would also be true of the well-known expanding target patterns near the time of their birth, when their radius is $O(\epsilon)$. But that event occurs in only brief intervals of time, when each ring is first formed. The corner layer, of course, represents a steady persistent structure.

Rotating spiral waves, to be discussed in this section, also represent steady structures near whose center the first two terms in (28) very probably are comparable to each other in magnitude, although that magnitude in this case is not $O(1)$.

There is considerable evidence, and wide consensus, that (1), where $D \leq O(\epsilon)$ (and for even larger values of D), have steady rotating solutions in the plane with an interface (also rotating, of course) in the shape of a spiral.

Structures of this type ("rotors," in Winfree's terminology) and their three-dimensional analogues are very commonly seen in the BZ reagent and various kinds of biological media ([Win72], [Win73], [Win74b], [Win78], [Win80], [Win84], [Zh], [Zy80a], [Zy80b], [Zy84], [KMi], [IKS]).

The evidence for rotor solutions of (1) is independent of these indications. Numerical experiments on this and similar equations are convincing and do reveal rotors. Moreover, we can infer indirectly that such things probably occur. For example, there is a convincing phase-plane argument in [Win74b], [Win78], [Win87]. In our singular perturbation context, the argument in §3C shows that twists occur naturally when, at some instant of time, there is a point on Γ where $v = v^*$ and v is strictly monotone along Γ there. These twists almost certainly develop into rotors. This argument was described in [Fi84a], [Fi85]. It also was explained there how these initial conditions could be set up by disrupting a traveling pulse solution or a traveling band of chemical activity such as those that occur in the expanding ring patterns. Then typically two rotors are born simultaneously when the two broken ends begin to twist and curl. This disruption corresponds very closely to a common way that rotors are produced in experiments.

Assuming that steady rotor solutions of (1) exist, the question remains as to how to construct them in a formal but asymptotic manner, in the spirit of this monograph. This question has many subquestions:

(a) Do the rotors have interfaces, as described in this chapter?

(b) What are the characteristic length scales for the outer solutions and for the layers?

(c) What is the characteristic period of rotation?

(d) Are the solutions unique?

(e) Is there a "core" region near the center, whose structure has characteristic spatial dimensions different from those elsewhere in the rotor?

Of course, we could ask the same questions about experimentally observed rotors in the BZ reagent, say, but the answers are generally inclusive, except for (d): rotors appear to be unique, with a unique period. As for (b), (c), and (d), the characteristic length, measured by the distance between adjacent spirally wrapped bands, is generally somewhat smaller than that of all target patterns in the same reagent; their oscillations are somewhat faster; and although the term "core" is often used to describe the region near the center, it does not appear to this observer (of photographs) that the characteristic dimensions are any different there than in the region away from the center.

It would be of great interest to discuss the qualitative characteristics of rotor solutions of (1); this would be another way to judge how good (1) is as a model for spatiotemporal dynamics of the BZ reagent. Moreover, and just as importantly, if it is judged to be a good model, then the analysis of (1) will provide much needed insight into the *reasons* why these structures are as they are.

For definiteness, let us take $D = \alpha\epsilon$, $\alpha = O(1)$; this choice is in accordance with the interpretation of u and v in (1) as concentrations of chemicals with comparable diffusivities [TF].

There have been two main theories in very recent years about the nature of rotor solutions of (1) when ϵ is small. One was developed by Fife [Fi84b], [Fi85], who argues that rotor solutions exist with uniform characteristic spatial dimensions throughout. This argument is discussed in the next section.

The other theory was developed by Keener and Tyson [KT] (see also [Ke86]). They envisage rotor solutions for small ϵ whose characteristic dimensions near the center (in a "core" region) are smaller than those away from that region. These ideas are discussed in §4C.

4B. Uniform rotors. We will think of steady rotors as being the longtime asymptotic states reached by the twist solutions described in §3C. Near their centers, those twist solutions have natural space and time scales given by (49). It seems reasonable to first look for steadily rotating solutions that have that same natural scaling everywhere. The reason is that as the interface Γ wraps around more and more, the tightly packed center, composed of these wrappings, expands. It may well be that the tightness of the packing, measured in terms of the normal distance between neighboring branches of Γ, remains the same. In that case the fully developed rotor, which occupies all space, will have uniform space scale. This is what we call a uniform rotor.

We therefore write the original equations (1) in terms of the rescaled quantities (49) and look for steady rotor solutions with angular velocity ω in the rescaled variables. At the same time, we note that in (50) it is shown that the controller variable v deviates from the neutral value v^* by the amount $O(\epsilon^{1/3})$; therefore we rescale that variable as well by setting

$$v = v^* + \epsilon^{1/3} w. \tag{51}$$

The functions u and w now must be of the form

$$u = u(r, \tilde{\theta} - \omega\tau), \qquad w = w(r, \tilde{\theta} - \omega\tau), \tag{52}$$

where $(r, \tilde{\theta})$ are polar coordinates in the ξ plane. We substitute (49), (51), and (52) into (1) and define

$$\delta = \epsilon^{1/3}$$

as our new small parameter. Using the subscript "θ" to denote differentiation with respect to the angular variable and Δ_ξ to denote the Laplacian in the variable ξ, we obtain

$$f(u, v^* + \delta w) = \delta^2(-\omega u_\theta - \Delta_\xi u), \tag{53a}$$

$$\alpha\Delta_\xi w + \omega w_\theta + g(u, v^* + \delta w). \tag{53b}$$

We look for solutions (52) of (53) which are periodic of period 2π, of course, in their second argument. The lowest order outer solution would satisfy

(54a) $$f(u_0, v^*) = 0,$$

(54b) $$\alpha\Delta_\xi w + \omega w_\theta + g(u_0, v^*) = 0.$$

As usual we select, as the solution of (54a),

(55) $$u_0 = h_\pm(v^*) \quad \text{in } \mathcal{D}_\pm,$$

hence from (54b),

(56) $$\alpha\Delta_\xi w + \omega w_\theta + G_\pm(v^*) = 0 \quad \text{in } \mathcal{D}_\pm.$$

These equations will govern the lowest order outer solution in $\mathcal{D}_\pm$. But we also need to consider the law of motion of Γ. As in the case of the interfaces examined in §1, this law of motion is found from the properties of the inner solution. However, we have to realize that for steadily rotating functions, the normal velocity of the interface is a function of the slope of the interface in polar coordinates, the polar distance r, and the angular velocity ω. The law of motion we are going to seek will involve a relation among the normal velocity at a given point on Γ, the value of v (hence w) at that point, and the curvature there. For these rotating solutions this translates into a relation among the value of w at a point, the angular velocity ω, the polar distance r, and the slope and curvature of the interface there. This latter relation will be obtained directly from an inner analysis; it will be called a *free boundary condition,* because the steady shape of Γ is a free boundary between $\mathcal{D}_+$ and $\mathcal{D}_-$.

We therefore proceed to study the inner problem and to discover the correct interface condition. Let the interface Γ be given, at least in part, by some function

$$\Gamma: \quad \tilde{\theta} = \gamma(r).$$

Since Γ might not exist near the center of rotation (see Fig. 4.9), it may

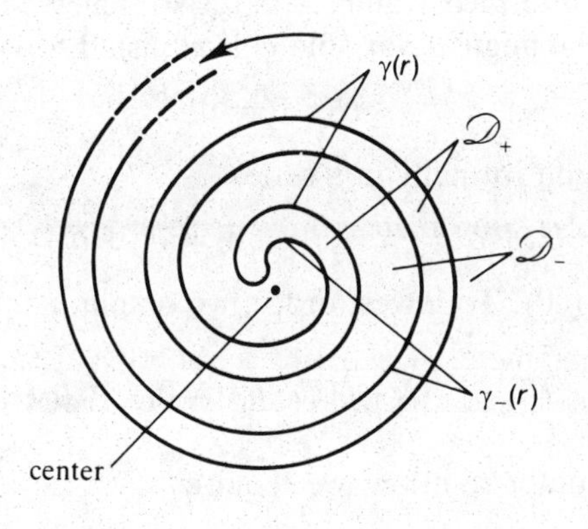

FIG. 4.9

happen that the function γ is defined only for $r \geq r_0$, say. For a one-armed rotor, moreover, there will be two interfaces for each $r > r_0$; an upjump Γ_+ and a downjump Γ_- (see Fig. 4.9). The upjump is characterized by the relation

$$\text{the point } (r, \gamma(r) \mp 0) \in \mathcal{D}_{\pm}. \tag{57}$$

We take the function γ to define the upjump branch Γ_+; downjumps will be discussed later. We use the shifted angular coordinate

$$\theta \equiv \tilde{\theta} - \gamma(r)$$

and the function $H(r) \equiv \gamma'(r)/r$, so that in the (r, θ) coordinate system, the Laplacian assumes the form

$$\Delta_H w \equiv \frac{1}{r} \partial_r (r \partial_r w) + \frac{1}{r^2} \partial_{\theta\theta} w - \frac{1}{r} \partial_r (r^2 H) \partial_\theta w - 2rH \partial_{r\theta} w + r^2 H^2 \partial_{\theta\theta} w. \tag{58}$$

To analyze the fine structure of the interface in the propagator variable, we have always used in the past a stretched coordinate ρ normal to Γ. However, in the present case we will use a stretched coordinate that is not normal to Γ, but rather is at an oblique angle. This choice leads to equivalent results and is more convenient. So we consider a point $(r, 0) \in \Gamma$. We "stretch" the angular variable, i.e., stretch in the direction perpendicular to the radius vector to that point, by defining

$$\phi = \frac{\theta}{\delta}, \quad W(r, \phi) \equiv w(r, \theta), \quad U(r, \phi) \equiv u(r, \theta). \tag{59}$$

With these variables, (53a) becomes

$$\begin{aligned} &\frac{1}{r^2} \partial_{\phi\phi} U + r^2 H^2 \partial_{\phi\phi} U + f(U, v^* + \delta W) \\ &\quad - \frac{\delta}{r} (r^2 H)' \partial_\phi U - 2\delta r H \partial_{r\phi} U + \delta \omega \partial_\phi U + \frac{\delta^2}{r} \partial_r (r \partial_r U) = 0. \end{aligned} \tag{60}$$

The location of the interface is defined in the stretched coordinates as the place where $U = 0$, and the angular variable ϕ is adjusted to vanish there:

$$U(r, 0) = 0.$$

Expand the solution formally in a series

$$U = U_0 + \delta U_1 + \cdots, \qquad W = W_0 + \delta W_1 + \cdots,$$

and substitute into (60). To lowest order, we obtain

$$\left(\frac{1}{r^2} + r^2 H^2\right) \partial_{\phi\phi} U_0 + f(U_0, v^*) = 0, \qquad U_0(r, 0) = 0.$$

To match with the outer solution, we require

$$U_0(r, \mp\infty) = h_{\pm}(v^*).$$

The solution is

$$U_0(r, \phi) = \psi(-p(r)\phi),$$

where $p(r) \equiv [r^{-2} + r^2 H(r)^2]^{-1/2}$ and $\psi(s)$ is the solution of

$$\psi'' + f(\psi, v^*) = 0, \quad \psi(0) = 0, \quad \psi(\pm\infty) = h_\pm(v^*),$$

introduced in (26) in Chap. 1. Considered as a traveling wave solution, its velocity is 0, which is in accordance with the definition of v^* in (9). So we call it a *stationary front solution.*

Carrying out the solution to the next order in δ, we obtain the equation

$$\text{(61)} \quad p^{-2}\partial_{\phi\phi} U_1 + f_u(U_0, v^*)U_1 + f_v(U_0, v^*)W_0 - \frac{1}{r}(r^2 H)'\partial_\phi U_0 - 2rH\partial_{r\phi} U_0 + \omega\partial_\phi U_0 = 0.$$

We also have that $\partial_{\phi\phi} W_0 = 0$, so that $W_0(r, \phi) \equiv w_0(r, 0)$ and v_0 is continuous.

We may write (61) in the form $LU_1 = F$, where L is the differential operator

$$LU \equiv p(r)^{-2}\partial_{\phi\phi} U + f_u(U_0, v^*)U$$

and F is a known function. This operator, with zero boundary conditions at $\pm\infty$, has a nullspace spanned by the single function $\chi(\phi) \equiv \psi'(-p(r)\phi)$. The next order problem's solvability condition is therefore that F be orthogonal in $L_2(-\infty, \infty)$ to that eigenfunction. Thus

$$\text{(62)} \quad \int_{-\infty}^{\infty} \chi(\phi)\left[f_v(U_0, v^*)w_0(r, 0) - \frac{1}{r}(r^2 H)'\partial_\phi U_0 - 2rH\partial_{r\phi} U_0 + \omega\partial_\phi U_0\right] d\phi = 0.$$

This can be solved for $w_0(r, 0)$ (we will omit the subscript "$_0$" in the following). We thus obtain the interface condition

$$\text{(63)} \qquad w(r, 0) = M\left(\left\{\frac{1}{r}[r^2 H(r)]' - \omega\right\}p(r) + rH(r)p'(r)\right),$$

M being a specific constant.

This was for the upjump interface. However, the same reasoning leads to the analogous free boundary condition on the downjump interface Γ_-. In that case, the left side of (63) would be replaced by $w(r, \gamma_-(r) - \gamma(r))$, where $\tilde{\theta} = \gamma_-(r)$ defines the downjump Γ_-, and on the right of (63) the functions p and H are replaced by p_- and H_-, corresponding to γ_- instead of γ.

The (lowest order) free boundary problem (**FBP**) for the determination of the rotor may now be stated.

FBP. Find constants ω, r_0, and β, and functions $w(r, \theta)$, $H(r)$, and $H_-(r)$, the latter two defined for $r \geq r_0$, such that

(i) (56) holds in $\mathcal{D}_\pm$;
(ii) w is 2π-periodic in θ;
(iii) $H(r)$ and $H_-(r) = (\beta/r) + O(r^{-2})$ $(r \to \infty)$;
(iv) (63) holds, together with the analogous condition on Γ_-; and
(v) the two curves $\theta = \gamma(r)$ and $\theta = \gamma_-(r)$ join smoothly at $r = r_0$.

Condition (iii) expresses the requirement that Γ be asymptotic to an Archimedian spiral for large r: $\gamma(r) \simeq \beta r$. Here β is the pitch of the asymptotic spiral. Substituting this asymptotic expression into (63) gives

$$w(\infty, 0) = -M\omega\beta^{-1}, \quad (64)$$

which relates, far from the center of the rotor, the angular velocity ω, the pitch β, and the value of w on Γ. This latter quantity, in turn, is related to the normal velocity c of Γ through (21), which now takes the form

$$c = Q(v^* + \delta w(\infty, 0)) = O(\delta), \quad (65)$$

since $Q(v^*) = 0$. In view of this, (64) can be seen as relating the normal velocity, the frequency, and the pitch of the interface far from its center. It is an automatic relation resulting from the geometry alone.

In addition to that relation, there is an independent one, called the *dispersion relation,* which can be obtained from (56) alone. Specifically, in the limit for large r where the curvature of Γ vanishes, (56) can be reformulated as a differential equation governing the properties of wave trains propagating in one direction. The independent variable in that equation is the coordinate in the direction of propagation. The periodic solutions of that equation can be found; they entail a relation between the frequency of the wave train (which is the same as our ω) and its velocity c. The relation so obtained is the asymptotic form of the dispersion relation for planar wave trains as $\epsilon \to 0$.

This shows that the correct scaling to discover that relation for small ϵ is the one we have used, namely (49). A study of the dispersion relation for a range of ϵ wider than this was given in [DKT]. That study verifies that the appropriate variables scale this way for very small ϵ.

Experimental evidence suggests that for each given ϵ, rotors exist for only a single value of ω, hence of β. Our conjecture is that there is a whole family of solutions of the **FBP**, one for each ω in some range, but that there is some selection mechanism that makes only one of them physically realizable. This type of selection phenomenon occurs in many areas of fluid mechanics; an example is the well-known Benard problem of convection in a fluid layer induced by buoyancy in the presence of a thermal gradient. If this is correct, then the selection mechanism would determine a unique frequency ω, and this would explain the uniqueness of observed rotors.

If $r_0 > 0$, we can say something about the character of the interface near the point $r = r_0$. Condition (v) above requires that $\gamma'(r)$, hence $\gamma''(r)$, become infinite as $r \downarrow r_0$. The right side of (61) must remain bounded, however, because the left side is. It turns out that this boundedness condition implies that the behavior of the function $\gamma(r)$ near r_0 is as follows:

$$\gamma(r) \simeq A(r - r_0)^{1/2} \quad (r \downarrow r_0). \quad (66)$$

The case $r_0 = 0$ arises when $G_+(v^*) = -G_-(v^*)$. This induces some symmetry in the **FBP**. Specifically, the symmetry implies that

$$w(r, \theta) = -w(r, \theta + \pi), \quad \gamma_-(r) = \gamma(r) + \pi, \quad r_0 = 0.$$

This case was discussed in [Fi84b], [Fi85]. Series expansions in powers of r for w and H can then be found; they are presumably valid for small r. They reveal, for instance, that $H(0)$ is proportional to ω. Similarly, series in powers of r^{-1} can be found, presumably valid for large r.

4C. Nonuniform rotors. In [KT], Keener and Tyson propose to construct rotating solutions of (1) for small ϵ in the following way. They assume that a solution exists with certain properties in a "core" region around the center with radius $O(\epsilon^{1/2})$, and different properties outside that region.

In the outside region, it is assumed that the value of v on Γ_+ (the upjump front) is constant, say equal to v_+, and on the wave back Γ_- (downjump) it is equal to v_-. Then in (28) the first term on the right will be constant; assume for the moment that it is given. In this region the last term in (28) may be neglected; we then obtain a relation between the normal velocity and the curvature. This translates into a differential equation for the shape of Γ_+ and Γ_-. As a boundary condition for the differential equation, Keener and Tyson impose what amounts to a no-flux condition at some unspecified distance r_1 from the center. Moreover, as $r \to \infty$, the angle $\tilde{\theta}$ being fixed, the boundary condition is that the rotor's structure should approach that of an outward moving periodic wave train. This exists only for v_+ and v_- satisfying $Q(v_+) = -Q(v_-)$, and only when this common value is related to ω by the dispersion relation mentioned in the previous section. This boundary value problem turns out to be solvable only if a certain second relation between ω and v_+ is satisfied. These two relations may be solved simultaneously to obtain a unique ω, hence a unique solution of the "outside" problem.

In the core region, the solution would be obtained by rescaling space by $r = \epsilon^{1/2}\hat{r}$. This results in an equation like (56), but with w replaced by v. In their rescaled problem, therefore, Keener and Tyson do not assume or take advantage of any small deviation of v from v^*, as we have done in (51). This is again a free boundary problem, and their free boundary condition is similar to (63). It is assumed that this problem has a solution with Γ_+ and Γ_- joining smoothly at some point (as we have assumed also) and matching with the solution in the outside region as $\hat{r} \to \infty$. This matching involves the value of v on $\Gamma_\pm$ approaching $v_\pm$.

One attractive feature of the Keener–Tyson concept is that the outside problem provides a unique value of ω. Therefore, if their idea is correct, it provides a straightforward way to calculate the unique frequency and velocity of the spiral rings far from the center. On the other hand, it involves a number of assumptions that appear to be no better justified than those in Fife's concept discussed in §4B.

4D. Conclusion. We now have two alternate proposals for the nature of steadily rotating solutions of (1). One of them says that such solutions exist with the property that the controller variable v differs from v^* everywhere by an amount of order $\epsilon^{1/3}$, and the spacing between adjacent winds of the spiral is everywhere $O(\epsilon^{2/3})$. Moreover, the period of rotation is $O(\epsilon^{1/3})$.

The other proposal is that the spacing is not uniform; the windings become tighter near the center. The value of v on Γ is approximately constant except in a core region of size $O(\epsilon^{1/2})$ in which it changes from v_+ to v_- as Γ changes smoothly from Γ_+ to Γ_-. The order of magnitude of v_+ and v_- and the spacing of the bands in the outside region are not clear; they depend on the dispersion relation [DKT] which they propose to use in a form that is valid not only in the limit $\epsilon \to 0$, but also for a wider range of values of ϵ.

Both proposals lead to mathematical problems that are free boundary problems, as yet unsolved even numerically, and which may or may not have solutions. The real question, however, is whether either of these proposals adequately describes rotor solutions of (1) for small ϵ. Obvious possibilities are that one or the other does, or that neither one does. Another possibility is that they both do, i.e., that at least two different kinds of solutions exist, described by these two concepts, and maybe in fact there is some continuous range of solutions between them, all for small ϵ. Still another possibility is that for small enough ϵ, only uniform solutions exist as described in §2, but that as ϵ increases (but still remains $\ll 1$), the nature of those solutions changes so that a separate core region emerges, as envisaged in [KT]. These are purely mathematical questions and will have to be settled by mathematical methods. The fact, for example, that one proposal entails an easy calculation of a unique ω is attractive, but has no bearing on the issue.

These problems pose challenges for the future. Another set of challenges is found in the various mathematical aspects of the dynamics of scroll waves [Win73] and their filament centers, the latter reminiscent of vortex filaments. This subject is gaining wide current interest. See the informative discussions in [Win87].

References

(*Following each reference is an indication of the sections to which it is relevant. Not all of these items are mentioned in the text.*)

[AB] N. Alikakos and P. Bates, *On the singular limit in a phase field model of phase transitions,* Ann. Inst. H. Poincaré, Anal. Non Linéaire, 5 (1988), pp. 1–38. (Chap. 1, §§2, 3.)

[ASh] N. D. Alikakos and K. C. Shaing, *On the singular limit for a class of problems modelling phase transitions,* SIAM J. Math. Anal., 18 (1987), pp. 1453–1462. (Chap. 1, §§2, 3.)

[ASi] N. Alikakos and H. C. Simpson, *A variational approach for a class of singular perturbation problems and applications,* Proc. Roy. Soc. Edinburgh, 107A (1987), pp. 27–42. (Chap. 1, §§2, 3.)

[AMP] S. B. Angenent, J. Mallet-Paret, and L. A. Peletier, *Stable transition layers in a semilinear boundary value problem,* J. Differential Equations, 67 (1987), pp. 212–242. (Chap. 1, §2.)

[AW75] D. G. Aronson and H. F. Weinberger, *Nonlinear diffusion in population genetics, combustion and nerve propagation,* in Proceedings of the Tulane Program in Partial Differential Equations and Related Topics, Lecture Notes in Mathematics 466, Springer-Verlag, Berlin, 1975, pp. 5–49. (Chap. 1, §2; Chap. 4, §1.)
[Basic properties of scalar nonlinear diffusive waves.]

[AW78] ———, *Multidimensional nonlinear diffusion arising in population genetics,* Adv. in Math., 30 (1978), pp. 33–76. (Chap. 1, §2; Chap. 4, §1.)
[Continuation of the preceding reference.]

[Av] J. Avrin, *Global existence for a model of electrophoretic separation,* SIAM J. Math. Anal., 19 (1988), pp. 520–527. (Chap. 3.)
[Existence for the initial value problem for 3 species electrophoresis; no electroneutrality assumption.]

[BZY] V. G. Babskii, M. Yu. Zhukov, and V. I. Yudovich, *Mathematical Theory of Electrophoresis—Application to Methods of Fractionation of Biopolymers,* Naukova Dumka, Kiev, 1983. (Chap. 3.)
[The most comprehensive physical and mathematical survey of electrophoresis available.]

[BZ] G. I. Barenblatt and Ya. B. Zel'dovich, *On the stability of propagating flames,* Prikl. Matem. i Mekh., 21 (1957). (Chap. 2, §2.)

[BNS83] H. Berestycki, B. Nicolaenko, and B. Scheurer, *Travelling wave solutions to reaction-diffusion systems modeling combustion,* in Nonlinear Partial Differential Equations, Contemporary Mathematics Vol. 17, Amer. Math. Soc., Providence, 1983, pp. 189–208. (Chap. 2, §1.)

[BNS85] ———, *Travelling wave solutions to combustion models and their singular limits,* SIAM J. Math. Anal., 16 (1985), pp. 1207–1242. (Chap. 2, §1.)

[BR] V. S. BERMAN AND IU. S. RIAZANTSEV, *Asymptotic analysis of stationary propagation of the front of a two-stage exothermic reaction in a gas,* J. Appl. Math. Mech. (PMM), 37 (1973), pp. 995–1004. (Chap. 2, §§1, 4.)

[BPMS] M. BIER, O. A. PALUSINSKI, R. A. MOSHER, AND D. A. SAVILLE, *Electrophoresis: Mathematical modeling and computer simulations,* Science, 219 (1983), pp. 1281–1287. (Chap. 3.)

[BS87] C. M. BRAUNER AND Cl. SCHMIDT-LAINÉ, *Existence of a solution to a certain plane premixed flame problem with two-step kinetics,* SIAM J. Math. Anal., 18 (1987), pp. 1406–1415. (Chap. 2, §1.).

[BS88] ———, *On the stability of singular traveling waves in some combustion problems,* in Mathematical Modeling in Combustion and Related Topics, NATO ASI Series 140, C. M. Brauner and Cl. Schmidt-Lainé, eds., M. Nijhoff, Boston, Dordrecht, Lancaster, 1988.

[BL82] J. BUCKMASTER AND G. S. S. LUDFORD, *Theory of Laminar Flames,* Cambridge Univ. Press, New York, 1982. (Chap. 2, §§1, 2.)
[This and the following are references for many aspects of flame theory.]

[BL83] ———, *Lectures on Mathematical Combustion,* CBMS-NSF Conference Series in Appl. Math. 43, Society for Industrial and Applied Mathematics, Philadelphia, 1983. (Chap. 2, §§1, 2.)

[Ca] G. CAGINALP, *An analysis of a phase field model of a free boundary,* Arch. Rat. Mech. Anal., 92 (1986), pp. 205–245. (Chap. 1, §3.)

[CF86a] G. CAGINALP AND P. C. FIFE, *Phase field methods for interfacial boundaries,* Phys. Rev. B, 33 (1986), pp. 7792–7794. (Chap. 1, §3.)

[CF86b] ———, *Higher order phase field models and detailed anisotropy,* Phys. Rev. B, 34 (1986), pp. 4940–4943. (Chap. 1, §3.)

[CF87a] ———, *Elliptic problems involving phase boundaries satisfying a curvature condition,* IMA J. Appl. Math., 38 (1987), pp. 195–217. (Chap. 1, §3.)

[CF87b] ———, *Phase field models of free boundary problems: exterior boundaries, higher order equations, and anisotropy,* in Structure and Dynamics of Partially Solidified Systems, D. E. Loper, ed., NATO ASI Series, M. Nijhoff, Boston, Dordrecht, Lancaster, 1987, pp. 131–140. (Chap. 1, §3.)

[CF88] ———, *Dynamics of layered interfaces arising from phase boundaries,* SIAM J. Appl. Math., 48 (1988), pp. 506–518. (Chap. 1, §§1, 3.)
[More details for Chap. 1, §3, and extensions of those results.]

[CH] G. CAGINALP AND S. HASTINGS, *Properties of some ordinary differential equations related to free boundary problems,* Proc. Roy. Soc. Edinburgh, 104 (1986), pp. 179–204. (Chap. 1, §3.)

[CM] G. CAGINALP AND B. MCLEOD, *The interior transition layer for an ordinary differential equation arising from solidification theory,* Quart. Appl. Math., 44 (1986), p. 155. (Chap. 1, §3.)

[CHi] J. W. CAHN AND J. E. HILLIARD, *Free energy of a nonuniform system* I. *Interfacial free energy,* J. Chem. Phys., 28 (1957), pp. 258–267. (Chap. 1, §3.)

[CCL] R. CASTEN, H. COHEN, AND P. LAGERSTROM, *Perturbation analysis of an approximation to Hodgkin–Huxley theory,* Quart. Appl. Math., 32 (1975), pp. 365–402. (Chap. 4, §1.)
[Possibly the first use of internal layer asymptotics in studying excitable media.]

[Cl] P. CLAVIN, *Dynamic behavior of premixed flame fronts in laminar and turbulent flows,* Progr. Energy Combust. Sci., 11 (1985), pp. 1–59. (Chap. 2, §§1, 2, 4.)
[A good review article.]

[CFN] P. CLAVIN, P. C. FIFE, AND B. NICOLAENKO, *Multiplicity and related phenomena for flames with competitive networks,* SIAM J. Appl. Math., 47 (1987), pp. 296–331. (Chap. 2, §§1, 2, 4.)
[Basic structure and stability analysis for several 2-reaction networks.]

[CL] P. CLAVIN AND A. LIÑAN, *Theory of gaseous combustion,* in Nonequilibrium Cooperative Phenomena in Physics and Related Fields, M. G. Velarde, ed., Plenum, New York, 1984, p. 291. (Chap. 2, §1.)

[CNR] D. S. COHEN, J. C. NEU, AND R. R. ROSALES, *Rotating spiral wave solutions of reaction-diffusion equations,* SIAM J. Appl. Math., 35 (1978), p. 536. (Chap. 4, §4.)

[De] Z. DEYL, ED., *Electrophoresis: A Survey of Techniques and Applications,* Elsevier, Amsterdam, 1979. (Chap. 3.)

[DKT] J. D. DOCKERY, J. P. KEENER, AND J. J. TYSON, *Dispersion of traveling waves in the Belousov–Zhabotinskii reaction,* Phys. D, in press. (Chap. 4, §1.)
[Analysis of the dispersion relation for traveling wave trains.]

[Ec] W. ECKHAUS, *Asymptotic Analysis of Singular Perturbation,* North-Holland, Amsterdam, 1979. (Chap. 1, §1.)

[EH] T. ERNEUX AND M. HERSCHKOWITZ-KAUFMAN, *Rotating waves as asymptotic solution of a model chemical reaction,* J. Chem. Phys., 66 (1977), p. 248. (Chap. 4, §4.)

[FO] D. FEINN AND P. ORTOLEVA, *Catastrophe and propagation in chemical reactions,* J. Chem. Phys., 67 (1977), p. 2119. (Chap. 4, §§1, 3.)

[FKN] R. J. FIELD, E. KORÖS, AND R. M. NOYES, *Oscillations in chemical systems,* II. *Thorough analysis of temporal oscillation in the bromate-cerium-malonic acid system,* J. Amer. Chem. Soc., 94 (1972), pp. 8649–8664. (Chap. 4, §§1, 2, 4.)

[Fi74] P. C. FIFE, *Transition layers in singular perturbation problems,* J. Differential Equations, 15 (1974), pp. 77–105. (Chap. 1, §2.)

[Fi76a] ———, *Boundary and interior transition layer phenomena for pairs of second order differential equations,* J. Math. Anal. Appl., 54 (1976), pp. 497–521. (Chap. 4, §1.)

[Fi76b] ———, *Singular perturbation and wave front techniques in reaction-diffusion problems,* in SIAM-AMS Proceedings, Symposium on Asymptotic Methods and Singular Perturbations, New York, 1976, pp. 23–49.
[Asymptotics for wave front solutions of various kinds of 2-equation systems.]

[Fi76c] ———, *Pattern formation in reacting and diffusing systems,* J. Chem. Phys., 64 (1976), pp. 854–864. (Chap. 4, §3.)
[Use of layer asymptotics to describe the formation and movement of patterns in a model system.]

[Fi77] ———, *Asymptotic analysis of reaction-diffusion wave fronts,* Rocky Mountain J. Math., 7 (1977), pp. 389–415. (Chap. 4, §1.)
[More details and extensions for some fronts discussed in [Fi76b].]

[Fi78] ———, *Asymptotic states for equations of reaction and diffusion,* Bull. Amer. Math. Soc., 84 (1978), pp. 693–726. (Chap. 1, §2; Chap. 4, §1.)
[Survey of long-time behavior and patterns for reaction-diffusion systems.]

[Fi79a] ———, *The bistable nonlinear diffusion equation: basic theory and some applications,* in Applied Nonlinear Analysis, V. Lakshmikantham, ed., Academic Press, New York, 1979, pp. 143–160.

[Fi79b] ———, *Long time behavior of solutions of bistable nonlinear diffusion equations,* Arch. Rat. Mech. Anal., 70 (1979), pp. 31–46. (Chap. 1, §§2, 3.)

[Fi79c] ———, *Mathematical Aspects of Reacting and Diffusing Systems,* Lecture Notes in Biomathematics 28, Springer-Verlag, New York, 1979. (Chap. 1, §2; Chap. 4, §1.)
[Survey of some methods useful for modeling and analyzing reaction-diffusion systems.]

[Fi81] ———, *On the question of the existence and nature of homogeneous-center target patterns in the Belousov–Zhabotinskii reagent,* in Analytical and Numerical Approaches to Asymptotic Problems in Analysis, O. Axelsson et al., eds., Mathematics Studies 47, North-Holland, Amsterdam, 1981, pp. 45–56. (Chap. 4, §§1, 2.)

[Fi82] ———, *Propagating fronts in reactive media,* in Nonlinear Problems, Present and Future, A. Bishop, D. Campbell, and B. Nicolaenko, eds., North-Holland, Amsterdam, 1982, pp. 267–285. (Chap. 1, §§2, 4; Chap. 4, §1.)

[Fi84a] ———, *Current topics in reaction-diffusion systems,* in Nonequilibrium Cooperative Phenomena in Physics and Related Fields, M. G. Velarde, ed., Plenum, New York, 1984, pp. 371–412. (Chap. 4, §§1, 2, 3, 4.)
[Description of the method in Chap. 4, §§1, 2.]

[Fi84b] P. C. FIFE, *Propagator-controller systems and chemical patterns,* in Nonequilibrium Dynamics in Chemical Systems, A. Pacault and C. Vidal, eds., Springer-Verlag, Berlin, New York, 1984, pp. 76–88. (Chap. 4, §§1, 2, 4.)
[Much the same as the following item. Here the theory of uniform rotors is given, including a mechanism by which they are generated.]

[Fi85] ———, *Understanding the patterns in the BZ reagent,* J. Statist. Phys., 39 (1985), pp. 687–703. (Chap. 4, §§1, 2, 4.)
[Generation and structure of spiral solutions.]

[Fi, pr] ———, *Modeling and analysis of hydrogen-oxygen flames,* 1986, preprint. (Chap. 2, §4.)

[Fi88] ———, *Modeling the chemistry in flames,* in Mathematical Modeling and Related Topics, NATO ASI Series 140, C. M. Brauner and Cl. Schmidt-Lainé, eds., M. Nijhoff, Boston, Dordrecht, Lancaster, 1988. (Chap. 2, §4.)
[Chap. 2, §4 is based on this.]

[FGe] P. C. FIFE AND X. GENG, *Solidification waves in the phase-field model,* in preparation. (Chap. 1, §3.)

[FGr] P. C. FIFE AND W. M. GREENLEE, *Interior transition layers for elliptic boundary value problems with a small parameter,* Uspekhi Mat. Nauka, 29 (1974), pp. 103–130. (Chap. 1, §2.)

[FH] P. C. FIFE AND L. HSIAO, *The generation and propagation of internal layers,* J. Nonlinear Anal. TMA, 12 (1988), pp. 19–41. (Chap. 1, §2.)

[FM] P. C. FIFE AND J. B. MCLEOD, *The approach of solutions of nonlinear diffusion equations to travelling front solutions,* Arch. Rational Mech. Anal., 65 (1977), pp. 335–361. Also, Bull. Amer. Math. Soc., 81 (1975), pp. 1075–1078. (Chap. 1, §§2, 3; Chap. 4, §1.)
[Global stability of wave solutions of the bistable equations.]

[FN82] P. C. FIFE AND B. NICOLAENKO, *The singular perturbation approach for flame theory with chain and competing reactions,* in Ordinary and Partial Differential Equations, W. N. Everitt and B. D. Sleeman, eds., Lecture Notes in Mathematics 964, Springer-Verlag, Berlin, 1982, pp. 232–250. (Chap. 2, §§1, 4.)
[The temperature plateau and nonuniqueness phenomena revealed and studied.]

[FN83] ———, *Asymptotic flame theory with complex chemistry,* in Nonlinear Partial Differential Equations, J. Smoller, ed., Contemporary Mathematics 17, Amer. Math. Soc., Providence, 1983, pp. 235–256. (Chap. 2, §4.)

[FN84a] ———, *How chemical structure determines spatial structure in flame profiles,* in Modelling of Patterns in Space and Time, W. Jager and J. D. Murray, eds., Lecture Notes in Biomathematics 55, Springer-Verlag, Berlin, 1984, pp. 73–86. (Chap. 2, §4.)
[The most complete version of the Fife–Nicolaenko approach to complex chemistry in flames.]

[FN84b] ———, *Flame fronts with complex chemical networks,* Phys. 12D (1984), pp. 182–197. (Chap. 2, §4.)

[FN86] ———, *Flame structure for reaction networks with chain branching and recombination,* in Lectures in Applied Mathematics Vol. 24, Amer. Math. Soc., Providence, 1986, pp. 273–292. (Chap. 2, §4.)

[FPS] P. C. FIFE, O. PALUSINSKI, AND Y. SU, *Electrophoretic traveling waves,* Trans. Amer. Math. Soc., in press. (Chap. 3.)

[Fis] R. A. FISHER, *The wave of advance of advantageous genes,* Ann. Eugenics, 7 (1937), pp. 355–369. (Chap. 4, §1.)

[Fit] R. FITZHUGH, *Mathematical models of excitation and propagation in nerves,* in Biological Engineering, H. P. Schwan, ed., McGraw-Hill, New York, 1969. (Chap. 4, §1.)

[FNH] H. FUJII, Y. NISHIURA, AND Y. HOSONO, *On the structure of multiple existence of stable stationary solutions in systems of reaction-diffusion equations,* Stud. Math. Appl., 18 (1986), pp. 159–219. (Chap. 4, §1.)
[Survey of extensive work, mainly by the authors, of patterns in a system of two equations.]

[Ga] M. GARBEY, *Quasilinear parabolic-hyperbolic singular perturbation problem: application to 2-phase flow in porous media equation,* manuscript. (Chap. 1, §4.)

[Ge] X. GENG, in preparation. (Chap. 3.)

[Gr76] J. M. GREENBERG, *Periodic solutions to reaction-diffusion equations,* SIAM J. Appl. Math., 30 (1976), pp. 199–205. (Chap. 4, §§1, 4.)

[Gr78] ———, *Axisymmetric time-periodic solutions of reaction-diffusion equations,* SIAM J. Appl. Math., 34 (1978), pp. 391–397. (Chap. 4, §§1, 4.)

[Gr80] ———, *Spiral waves for $\lambda - \omega$ systems,* SIAM J. Appl. Math., 39 (1980), pp. 301–309. (Chap. 4, §4.)

[GHH] J. M. GREENBERG, B. D. HASSARD, AND S. P. HASTINGS, *Pattern formation and periodic structures in systems modeled by reaction-diffusion equations,* Bull. Amer. Math. Soc., 84 (1978), pp. 1296–1327. (Chap. 4, §4.)

[GH] J. M. GREENBERG AND S. P. HASTINGS, *Spatial patterns for discrete models of diffusion in excitable media,* SIAM J. Appl. Math., 34 (1978), pp. 515–523. (Chap. 4, §§1, 4.)

[Hab] N. HABERMAN, *Note on the initial formation of shocks,* SIAM J. Appl. Math., 46 (1986), pp. 16–19. (Chap. 1, §4.)

[Hag81] P. HAGAN, *Target patterns in reaction-diffusion systems,* Adv. Appl. Math., 2 (1981), pp. 400–416. (Chap. 4, §2.)

[Hag82] ———, *Spiral waves in reaction-diffusion equations,* SIAM J. Appl. Math., 42 (1982), pp. 762–785. (Chap. 4, §4.)

[HaH] B. I. HAIKIN AND S. I. HUDYAEV, *On the nonuniqueness of the temperature and combustion velocity under the action of concurrent reactions,* Dokl. Akad. Nauk SSSR, 245 (1979), pp. 155–158. (Chap. 2, §4.)

[Has] S. P. HASTINGS, *Multiple traveling waves in a combustion model,* SIAM J. Math. Anal., 19 (1988), to appear. (Chap. 2, §§1, 4.)

[HLW86] S. P. HASTINGS, C. LU, AND Y-H. WAN, *A three dimensional shooting method as applied to a problem in combustion theory,* Phys. 19D (1986), pp. 301–306. (Chap. 2, §§1, 4.)

[HLW87] ———, *Existence of a travelling flame front in a model with no cold boundary difficulty,* SIAM J. Appl. Math., 47 (1987), pp. 1229–1240. (Chap. 2, §§1, 4.)

[HP83] S. P. HASTINGS AND A. POORE, *A nonlinear problem arising from combustion theory: Liñan's problem,* SIAM J. Math. Anal., 14 (1983), pp. 425–430. (Chap. 2, §§1, 4.)

[HP85] ———, *On Liñan's problem from combustion theory* II, SIAM J. Math. Anal., 16 (1985), pp. 331–340. (Chap. 2, §§1, 4.)

[He] S. HEINZE, *Traveling waves in combustion processes with complex chemical networks,* Trans. Amer. Math. Soc., in press (Chap. 1, §§1, 4.)
[Existence proof for a wide variety of steady flame profiles with general kinetics.]

[HH] P. C. HOHENBERG AND B. I. HALPERIN, *Theory of dynamic critical phenomena,* Rev. Modern Phys., 49 (1977), pp. 435–485. (Chap. 1, §3.)

[HK] L. N. HOWARD AND N. KOPELL, *Slowly varying waves and shock structures in reaction-diffusion equations,* Stud. Appl. Math., 56 (1977), pp. 99–145. (Chap. 4, §1.)

[HNZ] J. M. HYMAN, B. NICOLAENKO, AND S. ZALESKI, *Order and complexity in the Kuramoto–Sivashinsky model of weakly turbulent interfaces,* Phys. 23D (1986), pp. 265–292. (Chap. 2, §2.)

[IMN] H. IKEDA, M. MIMURA, AND Y. NISHIURA, *Global bifurcation phenomena of traveling wave solutions for some bistable reaction-diffusion systems,* Nonlinear Anal. TMA, in press. (Chap. 4, §1.)

[IKS] G. R. IVANITSKII, V. I. KRINSKY, AND E. E. SEL'KOV, *Mathematical Biophysics of a Cell,* Nauka, Moscow, 1978. (Chap. 4, §§1, 4.)

[Jo] C. R. K. JONES, *Stability of the travelling wave solution of the FitzHugh–Nagumo system,* Trans. Amer. Math. Soc., 286 (1984), pp. 431–469. (Chap. 4, §1.)
[Stability proof of the rapid pulse solution.]

[JC75] G. JOULIN AND P. CLAVIN, *Asymptotic analysis of a premixed laminar flame governed by a two-step reaction,* Combust. and Flame, 25 (1975), pp. 389–392. (Chap. 2, §§1, 4.)

[JC79] ———, *Linear stability analysis of nonadiabatic flames: diffusional-thermal model,* Combust. and Flame, 35 (1979), pp. 139–153. (Chap. 2, §2.)

[JLLPS] G. JOULIN, A. LIÑAN, G. S. S. LUDFORD, N. PETERS, AND C. SCHMIDT-LAINÉ, *Flames with chain branching/chain breaking kinetics*, SIAM J. Appl. Math., 45 (1985), pp. 420–434. (Chap. 2, §4.)

[Ka] YA. I. KANEL', *On the stabilization of solutions of the Cauchy problem for the equations arising in the theory of combustion*, Mat. Sb., 59 (1962), pp. 245–288. (Chap. 1, §2; Chap. 4, §1.)

[KL] A. KAPILA AND G. S. S. LUDFORD, *Two-step sequential reactions for large activation energies*, Combust. and Flame, 29 (1977), pp. 167–176. (Chap. 2, §§1, 4.)

[KG81] A. L. KAWCZYNSKI AND J. GORSKI, *A model of nonsustained leading center*, Bull. de L'Acad. Pol. des Sci., 24 (1981), pp. 261–268. (Chap. 4, §1.)

[KG, ip] ———, *Chemical models of non-sustained leading center*, Polish J. Chem., in press. (Chap. 4, §1.)

[Ke80] J. P. KEENER, *Waves in excitable media*, SIAM J. Appl. Math., 39 (1980), pp. 528–548. (Chap. 4, §§1, 4.)

[Ke86] ———, *A geometrical theory for spiral waves in excitable media*, SIAM J. Appl. Math., 46 (1986), pp. 1039–1056. (Chap. 4, §§3, 4.)

[KT] J. P. KEENER AND J. J. TYSON, *Spiral waves in the Belousov–Zhabotinskii reaction*, Phys. 21D (1986), pp. 307–324. (Chap. 4, §4.)
[Approach discussed in Chap. 4, §4.]

[Kog] S. KOGA, *Rotating spiral waves in reaction-diffusion systems—phase singularities of multi-armed spirals*, Progr. Theor. Phys., 67 (1982), pp. 164–178. (Chap. 4, §4.)

[Koh] F. KOHLRAUSCH, *Über Concentrations-Verschiebungen durch Electrolyse im Inneren von Lösungen and Lösungsgemischen*, Ann. Phys., Leipzig, 62 (1897), p. 209. (Chap. 3.)

[Kop81] N. KOPELL, *Target pattern solutions to reaction-diffusion equations in the presence of impurities*, Adv. Appl. Math., 2 (1981), pp. 389–399. (Chap. 4, §2.)
[$\lambda - \omega$ systems used; also in the following three papers.]

[Kop83] ———, *Forced and coupled oscillators in biological applications*, Proc. Internat. Congr. Math., Warsaw, 1983. (Chap. 4, §§1, 2, 4.)

[KH73] N. KOPELL AND L. N. HOWARD, *Plane wave solutions to reaction-diffusion equations*, Stud. Appl. Math., 52 (1973), pp. 291–328. (Chap. 4, §1.)

[KH81] ———, *Target patterns and spiral solutions to reaction-diffusion equations with more than one space dimension*, Adv. Appl. Math., 2 (1981), pp. 212–238. (Chap. 4, §§2, 4.)

[KA] V. I. KRINSKII AND K. I. AGLADZE, *Interaction of rotating waves in an active chemical medium*, Phys. 8D (1983), pp. 50–56. (Chap. 4, §4.)

[KMa] V. I. KRINSKII AND B. A. MALOMED, *Quasi-harmonic rotating waves in distributed active systems*, Phys. 9D (1983), pp. 81–95. (Chap. 4, §4.)

[KMP] V. I. KRINSKII, A. B. MEDVINSKII, AND A. V. PANFILOV, *The evolution of autowave spirals*, Matematika Kibernetika, 8 (1986), pp. 1–48. (Chap. 4, §4.)

[KMi] V. I. KRINSKII AND A. S. MIHAILOV, *Autowaves*, Znanie, Moscow, 1984. (Chap. 4, §§1, 2, 4.)

[Ku78] Y. KURAMOTO, *Diffusion-induced chaos in reactions systems*, Suppl. Progr. Theoret. Phys., 64 (1978), pp. 346–367. (Chap. 4, §4.)

[Ku84] ———, *Chemical Oscillations, Waves, and Turbulence*, Springer Series in Synergetics 19, Springer-Verlag, New York, 1984. (Chap. 4, §4.)

[KT75] Y. KURAMOTO AND T. TSUZUKI, *On the formation of dissipative structures in reaction-diffusion systems*, Progr. Theoret. Phys., 54 (1975), pp. 687–699. (Chap. 2, §2.)

[KT76] ———, *Persistent propagation of concentration waves in dissipative media far from thermal equilibrium*, Progr. Theoret. Phys., 55 (1976), pp. 356–369. (Chap. 2, §2.)

[La] J. S. LANGER, *Theory of the condensation point*, Ann. Physics, 41 (1967), pp. 108–157. (Chap. 1, §3.)

[Lin] X. B. LIN, *Shadowing lemma and singularly perturbed boundary value problems*, SIAM J. Appl. Math., to appear. (Chap. 1, §1; Chap. 4, §1.)

[Li] A. LIÑAN, *A theoretical analysis of premixed flame propagation with an isothermal chain reaction,* Tech. Report, Inst. Nac. Tech. Aerospacial "Esteban Terradas," Madrid, 1971. (Chap. 2, §§1, 4.)

[LR] A. LIÑAN AND M. RODRIGUEZ, *The ozone flame,* in Combustion and Nonlinear Phenomena, P. Clavin, B. Larrouturou, and P. Pelce, eds., Les Editions de Physique, 1985. (Chap. 2, §4.)

[LM] S. LUCKHAUS AND L. MODICA, *The Gibbs–Thompson relation with the gradient theory of phase transitions,* preprint. (Chap. 1, §3.)

[Lu] G. S. S. LUDFORD, *Low Mach number combustion,* in Reacting Flows: Combustion and Chemical Reactors, Part 1, G. S. S. Ludford, ed., Lectures in Applied Mathematics Vol. 24, Amer. Math. Soc., Providence, 1986, pp. 3–74. (Chap. 2, §§1, 2.)

[LP] G. S. S. LUDFORD AND N. PETERS, *Slowly varying flames with chain-branching–chain-breaking kinetics,* in Dynamics of Flames and Reactive Systems, J. R. Bowen, N. Manson, A. K. Oppenheim, and R. I. Soloukhin, eds., Vol. 95 of Progr. in Astronautics and Aeronautics, 1985, pp. 75–91. (Chap. 2, §§1, 4.)

[Mag] K. MAGINU, *Stability of periodic traveling wave solutions of a nerve conduction equation,* J. Math. Biol., 6 (1978), pp. 49–57. (Chap. 4, §1.)

[MM] S. B. MARGOLIS AND B. J. MATKOWSKY, *Nonlinear stability and bifurcation in the transition from laminar to turbulent flame propagation,* Comb. Sci. Tech., 34 (1983), pp. 45–77. (Chap. 2, §2.)

[Ma] M. MARION, *Mathematical study of a model with no ignition temperature for laminar plane flames,* in Reacting Flows: Combustion and Chemical Reactors, II, Lectures in Applied Mathematics Vol. 24, Amer. Math. Soc., Providence, 1986, pp. 239–252. (Chap. 2, §§1, 4.)

[MS] B. J. MATKOWSKY AND G. SIVASHINSKY, *On the stability of nonadiabatic flames,* SIAM J. Appl. Math., 40 (1981), pp. 255–260. (Chap. 2, §2.)

[MK] A. S. MIKHAILOV AND V. I. KRINSKY, *Rotating spiral waves in excitable media: the analytical results,* Phys. 9D (1983), pp. 346–371. (Chap. 4, §4.)

[MF] M. MIMURA AND P. C. FIFE, *A 3-component system of competition and diffusion,* Hiroshima Math. J., 16 (1986), pp. 189–207. (Chap. 4, §1.)
[Example of use of layer asymptotics in partitioning an ecological community.]

[MTH] M. MIMURA, M. TABATA, AND Y. HOSONO, *Multiple solutions of two-point boundary value problems of Neumann type with a small parameter,* SIAM J. Math. Anal., 11 (1980), pp. 613–631. (Chap. 4, §1.)
[An extension of [Fi76a].]

[Mo87a] L. MODICA, *The gradient theory of phase transitions and the minimal interface criterion,* Arch. Rat. Mech. Anal., 98 (1987), pp. 123–142. (Chap. 1, §3.)
[This and the following give the Γ-convergence approach to variational problems in the phase field model.]

[Mo87b] ———, *Gradient theory of phase transitions with boundary contact energy,* Ann. Inst. H. Poincaré, Anal. Non Linéaire, 4 (1987), pp. 487–512. (Chap. 1, §3.)

[MPH] S. C. MÜLLER, T. PLESSER, AND B. HESS, *The structure of the core of the spiral wave in the Belousov–Zhabotinskii reaction,* Science, 230 (1985), pp. 661–663. (Chap. 4, §4.)

[MS63] W. W. MULLINS AND R. F. SEKERKA, *Morphological stability of a particle growing by diffusion or heat flow,* J. Appl. Phys., 34 (1963), pp. 323–329. (Chap. 1, §3; Chap. 2, §2.)
[The first stability analysis of a solidification interface. In this subject, models usually have infinitely thin interfaces, as is the case with this and the next paper. A number of subsequent papers on this subject have been written, not referenced here.]

[MS64] ———, *Stability of a planar interface during solidification of a dilute binary alloy,* J. Appl. Phys., 35 (1964), pp. 444–451. (Chap. 1, §3; Chap. 2, §2.)

[NST] B. NICOLAENKO, B. SCHEURER, AND R. TEMAM, *Some global dynamical properties of the Kuramoto–Sivashinsky equation: nonlinear stability and attractors,* Phys. 16D (1985), pp. 155–183. (Chap. 2, §2.)

[Ni] Y. NISHIURA, *Stability analysis of traveling front solutions of reaction-diffusion systems—an application of the* SLEP *method,* Proc. IVth Internat. Conf. on Boundary and Interior Layers: Computational and Asymptotic Methods, Novosibirsk, 1986. (Chap. 4, §1.)

[NF85] Y. Nishiura and H. Fujii, *Stability theorem for singularly perturbed solutions to systems of reaction-diffusion equations,* Proc. Japan Academy, 61(A) (1985), pp. 329–332. (Chap. 4, §1.)

[NF86] ———, SLEP *method to the stability of singularly perturbed solutions with multiple internal transition layers in reaction-diffusion systems,* Proc. NATO Workshop "Dynamics of Infinite Dimensional Systems," J. Hale and S. N. Chow, eds., NATO ASI Series F-37, 1986, pp. 211–230. (Chap. 4, §1.)

[NF87] ———, *Stability of singularly perturbed solutions to systems of reaction-diffusion equations,* SIAM J. Math. Anal., 18 (1987), pp. 1726–1770. (Chap. 4, §1.)
[Stability proofs for layered solutions of some R-D systems.]

[NM] Y. Nishiura and M. Mimura, *Layer oscillations in reaction-diffusion systems,* SIAM J. Appl. Math., to appear. (Chap. 4, §1.)

[NF] R. M. Noyes and R. J. Field, *Oscillating chemical reactions,* Ann. Rev. Phys. Chem., 35 (1974), p. 95. (Chap. 4, §§1, 2.)

[OR73] P. Ortoleva and J. Ross, *Phase waves in oscillatory chemical reactions,* J. Chem. Phys., 58 (1973), pp. 5673–5680. (Chap. 4, §§1, 2.)

[OR74] ———, *On a variety of wave phenomena in chemical oscillations,* J. Chem. Phys., 60 (1974), pp. 5090–5107. (Chap. 4, §1.)

[OR75] ———, *Theory of propagation of discontinuities in kinetic systems with multiple time scales: fronts, front multiplicity, and pulses,* J. Chem. Phys., 63 (1975), pp. 3398–3408. (Chap. 4, §1.)

[OY] L. A. Ostrovskii and V. G. Yahno, *The formation of pulses in an excitable medium,* Biofizika, 20 (1975), pp. 489–493. (Chap. 4, §1.)
[First exploration of variable velocity fronts in excitable media.]

[PLP] G. Paczko, P. M. Lefdal, and N. Peters, *Reduced reaction schemes for methane, methanol and propane flames,* 21st Internat. Symp. on Combustion, The Combustion Institute, 1986. (Chap. 2, §4.)

[PL] J. Pelaez and A. Liñan, *Structure and stability of flames with two sequential reactions,* 1984, preprint. (Chap. 2, §§1, 2, 4.)
[Stability analysis for this scheme.]

[Pe] N. Peters, *Numerical and asymptotic analysis of systematically reduced reaction schemes for hydrocarbon flames,* in Numerical Simulation of Combustion Phenomena, R. Glowinski, B. Larrouturou, and R. Temam, eds., Lecture Notes in Physics 241, Springer-Verlag, Berlin, 1985, pp. 90–109. (Chap. 2, §4.)
[One important current example of attempts to reduce complex networks to workable proportions.]

[PK] N. Peters and R. J. Kee, *The computation of stretched laminar methane-air diffusion flames using a reduced four-step mechanism,* preprint. (Chap. 2, §4.)

[PS] N. Peters and M. D. Smooke, *Fluid dynamic-chemical interactions at the lean flammability limit,* Combustion and Flame, 60 (1985), pp. 171–182. (Chap. 2, §4.)

[PW] N. Peters and F. A. Williams, *The asymptotic structure of stoichiometric methane-air flames,* 1986, preprint. (Chap. 2, §4.)

[Pi] L. M. Pismen, *Multiscale propagation phenomena in reaction-diffusion systems,* J. Chem. Phys., 71 (1979), pp. 462–473. (Chap. 2, §1.)

[RK] J. Rinzel and J. B. Keller, *Travelling wave solutions of a nerve conduction equation,* Biophys. J., 13 (1973), pp. 1313–1337. (Chap. 4, §1.)

[RLW] B. Rogg, A. Liñan, and F. A. Williams, *Deflagration regimes of laminar flames modeled after the ozone decomposition flame,* Combustion and Flame, 65 (1986), pp. 79–101. (Chap. 2, §4.)

[RW] B. Rogg and I. S. Wichman, *Approach to asymptotic analysis of the ozone-decomposition flame,* Combustion and Flame, 62 (1985), pp. 271–293. (Chap. 2, §4.)

[Ro] C. D. Roten, *An algorithm for allocation and temperature, and its consequences for the chemistry of H_2–O_2 combustion,* in Mathematical Modeling in Combustion and Related Topics, NATO ASI Series 140, C. M. Brauner and Cl. Schmidt-Lainé, eds., M. Nijhoff, Boston, Dordrecht, Lancaster, 1988. (Chap. 2, §4.)

[RF] C. D. ROTEN AND P. C. FIFE, *Modelling and analysis of H_2–O_2 combustion between* 1100° *and* 1600°, preprint. (Chap. 2, §4.)

[RSK] J. RUBENSTEIN, P. STERNBERG, AND J. B. KELLER, *Fast reaction, slow diffusion, and curve shortening,* preprint. (Chap. 1, §2; Chap. 4, §1.)
[Gradient systems analogous to the equation (Chap. 1, equation (24)) in space are studied from the point of view of interfaces forming and evolving.]

[SaP] D. A. SAVILLE AND O. A. PALUSINSKI, *Theory of electrophoretic separations, Part* 1 *and Part* 2, AIChE J., 32 (1986). (Chap. 3.)

[SS] Cl. SCHMIDT-LAINÉ AND D. SERRE, *Etude de stabilité d'un système non linéaire de dimension* 4 *en combustion et generalisation a une classe de problèmes homogènes de degre* 2, Phys. 12D (1986), pp. 42–60. (Chap. 2, §§1, 2.)

[Se] M. SERMANGE, *Contributions to the numerical analysis of laminar stationary flames,* Lecture Notes in Physics 241, Springer-Verlag, Berlin, 1985, pp. 375–388. (Chap. 1, §1.)

[SeP] K. SESHADRI AND N. PETERS, *The influence of stretch on a premixed flame with two-step kinetics,* Comb. Sci. Tech., 33 (1983), pp. 35–63. (Chap. 1, §§1, 4.)

[Sh] K. SHOWALTER, *Trigger waves in the acidic bromate oxidation of ferroin,* J. Phys. Chem., 85 (1981), pp. 440–447. (Chap. 4, §1.)

[Si77a] G. SIVASHINSKY, *Diffusional-thermal theory of cellular flames,* Comb. Sci. Tech., 15 (1977), pp. 137–145. (Chap. 2, §2.)

[Si77b] ———, *Nonlinear analysis of hydrodynamic instability in laminar flames, Part* I, *Derivation of basic equations,* Acta Astronautica, 4 (1977), pp. 1177–1206. (Chap. 2, §2.)

[Si80] ———, *On flame propagation under conditions of stoichiometry,* SIAM J. Appl. Math., 39 (1980), pp. 67–82. (Chap. 2, §2.)

[SRRD] M. D. SMOOKE, H. RABITZ, Y. REUVEN, AND F. L. DRYER, *Application of sensitivity analysis to premixed hydrogen-air flames,* preprint. (Chap. 2, §4.)

[Su] Y. SU, *Qualitative analysis of isoelectric focussing: a single buffer and one sample model,* preprint. (Chap. 3.)

[SPF] Y. SU, O. A. PALUSINSKI, AND P. C. FIFE, *Isotachophoresis: Analysis and computation of the structure of the ionic species interface,* J. Chromatography, (1987), pp. 77–85. (Chap. 3.)

[SO1] R. SULTAN AND P. ORTOLEVA, *Rotating waves in reaction-diffusion systems with folded slow manifolds,* J. Chem. Phys., 84 (1986), pp. 6781–6789. (Chap. 4, §4.)
[They treat the case when the control variable diffuses faster than the propagator variable.]

[SO2] ———, *Static reaction-diffusion structures in folded slow manifold systems,* J. Chem. Phys., 85 (1986), pp. 5068–5075. (Chap. 4, §3.)

[Te, pre1] D. TERMAN, *The asymptotic stability of a traveling wave solution arising from a combustion model,* preprint. (Chap. 2, §§1, 2.)

[Te86] ———, *An application of the Conley index to combustion,* Proc. NATO Workshop "Dynamics of Infinite Dimensional Systems," J. Hale and S. N. Chow, eds., NATO ASI Series F-37, 1986. (Chap. 2, §§1, 4.)

[Te88] ———, *Connection problems arising from nonlinear diffusion equations,* in Microprogram on Nonlinear Diffusion Equations and Their Equilibrium States, W.-N. Ni, L. A. Peletier, and J. Serrin, eds., Springer-Verlag, New York, Berlin, 1988. (Chap. 2, §§1, 4.)

[Te, ip] ———, *Traveling wave solutions arising from a two-step combustion model,* SIAM J. Math. Anal., to appear. (Chap. 2, §§1, 4.)

[Th] W. THORMANN, *Principles of isotachophoresis and dynamics of the isotachophoretic separation of two components,* Separation Sci. Tech., 19 (1984), pp. 456–467. (Chap. 3.)

[TM] W. THORMANN AND R. A. MOSHER, *Recent developments in isotachophoresis,* in Isotachophoresis '86, VCH Verlagsgesellschaft, Weinheim, 1986, pp. 133–145. (Chap. 3.)

[Tr] W. C. TROY, *Mathematical modeling of excitable media in neurobiology and chemistry,* in Periodicities in Chemistry and Biology, Adv. in Chem. Education, Academic Press, New York, 1978. (Chap. 4, §1.)

[TNM] T. TSUJIKAWA, T. NAGAI, M. MIMURA, R. KOBAYASHI, AND H. IKEDA, *C_0-semigroup approach to stability for traveling pulse solutions of the FitzHugh–Nagumo equations,* preprint. (Chap. 4, §1.)

[Ty76] J. TYSON, *The Belousov–Zhabotinskii Reaction,* Lecture Notes in Biomathematics 10, Springer-Verlag, Berlin, 1976. (Chap. 4, §§1, 2, 4.)

[Ty79] ———, *Oscillations, bistability, and echo waves in models of the Belousov–Zhabotinskii reaction,* Ann. New York Acad. Sci., 36 (1979), pp. 279–295. (Chap. 4, §§1, 2, 4).

[Ty81] ———, *On scaling the Oregonator equations,* in Nonlinear Phenomena in Chemical Dynamics, C. Vidal and A. Pacault, eds., Springer-Verlag, New York, 1981, pp. 222–227. (Chap. 4, §§1, 2, 4.)

[Ty82] ———, *On scaling and reducing the Field–Korös–Noyes mechanism of the Belousov–Zhabotinskii reaction,* J. Phys. Chem., 86 (1982), pp. 3006–3012. (Chap. 4, §§1, 2, 4.)

[Ty85] ———, *A quantitative account of oscillations, bistability, and traveling waves in the Belousov–Zhabotinskii reaction,* in Oscillations and Traveling Waves in Chemical Systems, R. J. Field and M. Burger, eds., John Wiley, New York, 1985, pp. 93–144. (Chap. 4, §§1, 2, 4.)

[Ty, pre] ———, *Singular perturbation theory of target patterns in the Belousov–Zhabotinskii reaction,* preprint. (Chap. 4, §2.)
[It is pointed out that recent experiments corroborate the Tyson–Fife theory [TF].]

[TF] J. TYSON AND P. C. FIFE, *Target patterns in a realistic model of the Belousov–Zhabotinskii reaction,* J. Chem. Phys., 73 (1980), pp. 2224–2237. (Chap. 4, §§1, 2.)
[Use of layer asymptotics to model and study *BZ* targets.]

[VB] A. V. VASILEVA AND V. F. BUTUSOV, *Asymptotic Expansion of Solutions of Singular Perturbation Equations,* Nauka, Moscow, 1973. (Chap. 1, §1.)

[Wag] D. WAGNER, *Premixed laminar flames as travelling waves,* in Reacting Flows: Combustion and Chemical Reactors, II, Lectures in Applied Mathematics Vol. 24, Amer. Math. Soc., Providence, 1986, pp. 229–238. (Chap. 2, §1.)

[Wa] J. WARNATZ, *Chemistry of hydrocarbon combustion,* in Combustion and Nonlinear Phenomena, Les Editions de Physique, 1985. (Chap. 2, §4.)

[WD] C. K. WESTBROOK AND F. L. DRYER, *Chemical kinetics modeling of hydrocarbon combustion,* Progress in Energy and Combustion Science, 1983. (Chap. 2, §4.)

[Wil] F. A. WILLIAMS, *Combustion Theory,* Addison-Wesley, Reading, MA, 1965. (Chap. 2.)
[An excellent standard reference.]

[Win72] A. T. WINFREE, *Spiral waves of chemical activity,* Science, 175 (1972), pp. 634–636. (Chap. 4, §4.)

[Win73] ———, *Scroll-shaped waves of chemical activity in three dimensions,* Science, 181 (1973), pp. 937–939. (Chap. 4, §4.)

[Win74a] ———, *Wavelike activity in biological and chemical media,* in Mathematical Problems in Biology, P. van den Driessche, ed., Springer-Verlag, Berlin, 1974, pp. 241–260. (Chap. 4, §4.)

[Win74b] ———, *Rotating solutions of reaction-diffusion equations in simply-connected media,* in SIAM-AMS Proc. Vol. 8, Amer. Math. Soc., Providence, 1974, pp. 13–31. (Chap. 4, §4.)

[Win78] ———, *Stably rotating patterns of reaction and diffusion,* in Theoretical Chemistry Vol. 4, H. Eyring and D. Henderson, eds., Academic Press, New York, 1978, pp. 1–51. (Chap. 4, §4.)

[Win80] ———, *The geometry of biological time,* in Biomathematics Vol. 8, Springer-Verlag, New York, 1980. (Chap. 4, §§1, 2, 4.)
[A fascinating account of many of the authors' ideas relating to phase resetting, patterns in distributed systems, etc.]

[Win81] ———, *The rotor as a phase singularity of reaction-diffusion problems and its possible role in sudden cardiac death,* in Nonlinear Phenomena in Chemical Dynamics, C. Vidal and A. Pacault, eds., Springer-Verlag, Berlin, 1981, pp. 156–159. (Chap. 4, §4.)

[Win84] ———, *Wavefront geometry in excitable media: organizing centers,* Phys. 12D (1984), pp. 321–332. (Chap. 4, §4.)

[Win87] ———, *When Time Breaks Down,* Princeton University Press, Princeton, NJ, 1987.
[Another popular account of many things related to Chap. 4.]

[ZK] A. N. ZAIKIN AND A. L. KAWCZYNSKI, *Spatial effects in active chemical systems.* I. *Model of leading center,* J. Non-Equilib. Thermodynamics (1977), pp. 39–48. (Chap. 4, §§2, 4.)

[Ze] YA. B. ZEL'DOVICH, *On the theory of flame propagation,* Zh. Fiz. Khim., 22 (1948), p. 27. (Chap. 2, §1.)

[ZBLM] YA. B. ZEL'DOVICH, G. I. BARENBLATT, B. LIBROVICH, AND G. M. MAHVILADZE, *The Mathematical Theory of Combustion and Explosions,* Consultants Bureau (Translation), New York, London, 1985. (Chap. 2.)
[Contains, among other things, a comprehensive account of flame theory, including the effects of complex reactions.]

[ZF1] YA. B. ZEL'DOVICH AND D. A. FRANK-KAMENETSKII, *Theory of uniform flame propagation,* Dokl. Akad. Nauk SSSR, 19 (1938), p. 693. (Chap. 2, §1.)

[ZF2] ———, *The theory of thermal propagation of flames,* Zh. Fiz. Khim., 12 (1938), p. 100. (Chap. 2, §1.)

[Zh] A. M. ZHABOTINSKII, *Concentrational Self-Oscillations,* Nauka, Moscow, 1974. (Chap. 4.)

[ZZ] A. M. ZHABOTINSKII AND A. N. ZAIKIN, *Concentration wave propagation in two dimensional liquid-phase self-oscillating systems,* Nature, 225 (1970), pp. 535–537. (Chap. 4, §§1, 2.)

[ZZ] ———, *Auto-wave processes in a distributed chemical system,* J. Theoret. Biol., 40 (1973), pp. 45–61. (Chap. 4.)

[Zy80a] V. S. ZYKOV, *The kinematics of stationary circulation in an excitable medium,* Biophysics, 25 (1980), pp. 329–333. (Chap. 4, §§3, 4.)

[Zy80b] ———, *Analytic estimates for the dependence of the velocity of an excitation wave in a two dimensional excitable medium on the curvature of its front,* Biophysics, 25 (1980), pp. 906–911. (Chap. 4, §§3, 4.)

[Zy84] ———, *Modeling Wave Processes in Excitable Media,* Nauka, Moscow, 1984. (Chap. 4, especially §4.)
[Contains material related to that in Chap. 4, §§1, 2, 4. A comprehensive account of experiments, simulations, and modeling. Also the kinematics of spiral wave fronts, including the curvature relation. Translation by A. Winfree, Manchester Univ. Press, Manchester, to appear soon. It will contain extra material and an expanded bibliography.]